DATE DUE

GAYLORD			PRINTED IN U.S.A.

Before Big Science

Twayne's History of Science and Society Series

Mordechai Feingold, General Editor
Virginia Polytechnic Institute

Before Big Science

The Pursuit of Modern Chemistry and Physics 1800–1940

Mary Jo Nye

Twayne Publishers
An Imprint of Simon & Schuster Macmillan
New York

Prentice Hall International
London • Mexico City • New Delhi • Singapore • Sydney • Toronto

Before Big Science: The Pursuit of Modern Chemistry and Physics 1800–1940
Mary Jo Nye

Twayne's History of Science and Society Series, No. 1

Twayne Publishers
An Imprint of Simon & Schuster Macmillan
1633 Broadway
New York, New York 10019

Library of Congress Cataloging-in-Publication Data
Nye, Mary Jo.
Before big science : the pursuit of modern chemistry and physics 1800–1940 / Mary Jo Nye.
p. cm.—(The history of science and society ; 1)
Includes bibliographical references and index.
ISBN 0-8057-9512-X (cloth)
1. Chemistry—History—19th century. 2. Chemistry—History—20th century. 3. Physics—History—19th century. 4. Physics—History—20th century. I. Title. II. Series.
OD11.N94 1996
530′.09′034—dc20 96-22963
CIP

The paper used in this publication meets the minimum requirements of American National Standard for Information Sciences—Permanence of Paper for Printed Library Materials. ANSI Z39.48–1984. ♾™

To Bob and Lesley,
and to my parents Mildred Heath Mann and Joe Allen Mann

Contents

Foreword

Twayne's History of Science and Society series is designed to offer students as well as educated lay readers a comprehensive survey of Western science from ancient Greece to the present. Specialists, for their part, will find here accounts of the latest research along with the distinctive outlook of experts on their respective topics. The series comprises an interlocking whole that examines the totality of western scientific experience within its broader social and cultural context. While each volume will present a concise study of the nature and scope of the scientific enterprise during a particular period, several volumes will often be grouped together to treat related periods and to create a unifying narrative line. In this way the series hopes to avoid the limitations inherent in surveys that adhere to rigid periodization and strict boundaries between disciplines.

Mary Jo Nye's *Before Big Science* launches the series with a forceful account of the maturation of the fields of physics and chemistry during the nineteenth and first half of the twentieth centuries, a period that witnessed a spate of conceptual and experimental discoveries that transformed our understanding of the natural world. It was also during this period that physics and chemistry were reshuffled and their spheres of activity reconstituted. Concomitant with this transformation of the field was the resurgence of the universities as the main sites of scientific activity. Major research facilities were established there on an unprecedented scale and the concept of the teaching laboratory was introduced. Nye describes the rapid process of professionalization that accompanied these institutional changes and effectively put an end to the ideal of the "gentleman" practitioner that had reigned during the early modern period. It was also this period that witnessed improved experimental techniques, higher levels of precision, and near universal consensus over standardization, all of which offered scientists a greater scope for investigation. Another by-product of the enormous growth of the scientific community was an exponential increase in cost, which in turn forced government to underwrite a growing share of the expenditure. Finally, the period saw the internationalization of scientific activity and emergence of the United States, in particular, as a leading force.

Writing with authority and verve, Nye provides a cogent introduction to these and other aspects of the growth of modern physics and chemistry, one that will prove indispensable for anyone interested in understanding the rise of "Big Science."

Preface and Acknowledgments

In writing this history, I have had very much in mind the undergraduates and beginning graduate students whom I taught for twenty-five years in survey courses in the history of science at the University of Oklahoma. Their interests and their observations have informed this history. I also have had in mind my daughter, who was a liberal-arts college undergraduate when I began this book, as well as friends and family who are general readers. I hope that this book will be interesting and informative for science students and for scientists who enjoy reading about the histories of their fields.

In writing this volume, I have depended on general histories of science from antiquity to the present, histories of physics and histories of chemistry, biographical studies, institutional histories, and monographs written by historians, philosophers, sociologists, and scientists. I have drawn from articles and books written by the historical figures whose stories are recounted here. I have incorporated ideas and arguments that I have used in lectures over the course of many years, as well as some of the work of my students, including the dissertations of my students Jun Fudano, Yasu Furukawa, and Michael N. Keas.

Because this book does not contain the great number of footnotes customary in my usual published work, I have not been able to acknowledge or cite every author to whom credit is due. I ask the reader to take seriously the origins of this book in the sources cited in the bibliographical essay.

For their specific criticisms and suggestions, and for the help or inspiration they offered in various ways, I wish to thank my former students Kuangtai-Hsu, Mark A. Eddy, and Lynne A. Williams; my present student, J. Christopher Jolly; my daughter, Lesley N. Nye; and my friends and colleagues Diana K. Barkan, Jed Z. Buchwald, Joseph S. Fruton, Frederic L. Holmes, Daniel J. Kevles, Alan J. Rocke, and Niall Caldwell and Liba Taub. Mordechai Feingold enlisted me to write this book. I am grateful for the gentle pressure that he and Lesley Polliner put on me to complete it, as well as for the skillful copyediting by James Waller and the production and editorial oversight by Andrew Libby and Margaret Dornfeld. My greatest debt, as always, is owed to Robert A. Nye, for critical reading, insightful advice, and necessary encouragement.

For courteous assistance with photographs and permissions to publish, I gratefully acknowledge the Edgar Fahs Smith Collection at the University of Pennsylvania; Cambridge University Press; MIT Press; University of Wisconsin Press; the Royal Society of Chemistry; Godfrey Argent Photography Ltd.; the National Portrait Gallery, London; the Royal Society, London; the Manchester Literary and Philosophical Society; the History of Science Collections, University of Oklahoma Libraries; and the Special Collections (Ava Helen and Linus Pauling Papers) at Oregon State University, as well as the late Francis Perrin, Mme. Colette Grignard, and Dr. Keith U. Ingold.

Support from the Thomas Hart and Mary Jones Horning Endowment at Oregon State University and a by-fellowship at Churchill College at the University of Cambridge enabled me to complete this book. It was begun with support from the George Lynn Cross Professors' research fund at the University of Oklahoma, for which I also am grateful.

Introduction: Modern Science and Big Science

There is much talk in the late twentieth century about the giant scale of things: big government, big business, big science. In 1961, Alvin Weinberg, the director of the Oak Ridge National Laboratory in eastern Tennessee, wrote an influential commentary in *Science*, the weekly publication of the American Association for the Advancement of Science, on the state of high-energy physics in the United States.

Coining the phrase "Big Science," which he wrote with initial capitals, Weinberg expressed the view that giant-scale science was a stage in the development of science. The sociologist Derek J. de Solla Price had already plotted an exponential growth curve in 1956 showing that the sheer number of scientists and scientific papers doubled in size about every fifteen years after the seventeenth century, although Price inferred that the curve was beginning to level off in the late 1950s.

The 1960s period of big science was not, in Weinberg's opinion, the ideal outcome for the history of the sciences. In particular, Weinberg judged that the large sums of money spent on contemporary high-energy physics research were of fairly low value in contributing to the general human good. In his *Science* essay, Weinberg recalled for his readers the building of the ancient Egyptian pyramids and the medieval European cathedrals, enterprises that in their own times commanded exorbitant efforts from engineers, artists, laborers, and governments. He worried, "We must not allow ourselves, by shortsighted seeking after fragile monuments of Big Science, to be diverted from our real purpose, which is the enriching and broadening of human life."[1]

Historian Daniel J. Kevles noted in his history of the American physics community that Weinberg practiced what he preached. His own management of the Oak Ridge facility was one that directed some of the national laboratory's research effort to such socially useful concerns as cheap energy sources, desalination, and environmental problems.

Of course big science was not entirely new. To be sure, the coordination of hundreds and even thousands of individuals into group efforts across industry, government, military services, and the universities reached a new scale of personnel and funding levels as a consequence of World War II. Highly visible national and multinational particle accelerators, rockets, and space vehicles resulted after the war, as well as huge research and development programs in chemical products including synthetic materials and pharmaceuticals.

Yet already before 1940, Ernest Lawrence's cyclotron program at the University of California at Berkeley operated on the basis of an array of funding sources, with 22 percent of its capital and operating budget from the federal government, along with 40 percent from the state of California and 38 percent from private philanthropic foundations. By the 1930s scientists were complaining that administrative and committee work was taking too much time and that senior scientists had no time to do research. As early as 1900 some university professors and townspeople were complaining that some parts of universities were beginning to look like factories, as huge laboratory buildings were equipped with electrical generators, enormous magnets, vacuum pumps, chemicals, and heavy machinery.

The bigness of Big Science, as well as its undergirding theoretical and organizational structures, had clear origins in its recent history. Modern science was well established before the outbreak of World War II, and there emerged many important continuities, as well as significant disjunctions, between the history of chemistry and physics before and after the war.

This book studies the individuals, institutions, and ideas that turned the late eighteenth-century traditions of natural philosophy, natural history, and chemical philosophy into the twentieth-century disciplines of chemistry and physics that are familiar to contemporary students and readers of science.

Aspects of what we call chemistry and physics originally were taught in medieval and Renaissance universities as part of the curriculum in medicine, pharmacy, philosophy, astronomy, and mathematics. During the early decades of the 1800s the subject matter of physics and chemistry was found in lecture courses in many universities throughout Europe. Only in the mid-nineteenth century, however, did the words *chemistry* and *physics* acquire well-defined disciplinary meanings that are very much like their counterparts in the twentieth century.

Although early nineteenth-century chemistry continued to be associated with the traditions of natural history (botany, zoology, geology, mineralogy) and natural philosophy (mathematical and experimental physics, also known as mechanical philosophy), by mid-century chemistry and physics were coming into their own as "physical sciences" distinct from the moral and biological sciences.

In the first decades of the 1800s, chemists' principal goals lay in isolating and describing elementary substances and their properties, determining the elements' proportions in natural and synthesized substances, and identifying typical combinations of elements that function as chemically active radicals or groups. Chemists toyed with the idea that the origin of chemical force is electrical or gravitational in nature, but they did not overly concern themselves with the nature of chemical force or with possible mechanisms of combination. Rather, they laid emphasis on classifying simple and complex substances and in working out the kinds of relationships that resulted, in the late 1860s, in the periodic table of the elements devised by Dmitry Mendeleyev.

In contrast, physicists studied mechanics, optics, and acoustics, employing first geometry and then the analytic methods of integral and differential calculus in their treatments of these subjects. By the middle decades of the nineteenth century physicists were adding to their domain some experimental fields—especially heat and electricity—that they had previously shared with chemists but that chemists by and large now conceded to them. This concession on the part of chemists was largely due to the rapid growth of what came to be called "organic chemistry," which encompassed pharmacy, animal chemistry, agricultural chemistry, the chemistry of dyes, and the industrial chemistry of organic synthesis.

Organic chemistry developed a program of study, a language of discourse, and a system of explanation that was foreign to the practitioners of an earlier general chemistry, which had shared with natural philosophers and physicists a concern for corpuscular points, Newtonian forces, and subtle fluids. Further, as natural philosophy and physics became more and more mathematical, chemical philosophy and chemistry demonstrated very little need of mathematics beyond the simplest calculations.

The chemical laboratory became a vehicle for the teaching and training of a fairly large number of students. The laboratory was a different kind of scientific institution from the philosopher's lecture theater (where he gave public demonstrations) or the academic chemist's private rooms (for experimentation and testing). The development of the teaching laboratory gradually changed the character of natural philosophy, or physics, after the mid-nineteenth century, even as it had first transformed chemical philosophy. Not just experimentation but *precise* measurement and quantification using standardized and finely tuned (if not necessarily finely crafted) instruments, were to become rigorous requisites of physics as well as chemistry.

By the end of the nineteenth century, many physical and chemical laboratories looked like factories and were turning out data in a mechanical fashion. University science faculties required separate buildings, and the split between the "humanities" and the "sciences" became a division in physical as well as epistemological space. As noted earlier, by 1940, university-associated laboratories such as Ernest O. Lawrence's laboratory at the University of California at Berkeley were practically indistinguishable from factories or industrial laboratories to the casual observer.

Today's forms of scientific education and research evolved out of the expansion and reform of universities in the nineteenth century, which created not only the laboratory but also the seminar, the colloquium, the research institute, and schools of applied and engineering science independent from the military and technical schools of the eighteenth-century state. Increasing specialization within the teaching of the sciences dovetailed with increasing professionalization of the roles formerly held by generalist savants and scientific amateurs. By the end of the nineteenth century an array of new scientific societies, with distinct membership lists and published journals, proliferated alongside the older academies and philosophical societies. The physical sciences came to have great prestige, symbolized in the Nobel Prizes.

Diplomatic and cultural historians often have claimed that in important respects, the nineteenth century ended with World War I, not the year 1900. However, not a great deal changed in the everyday scientific life of scientists after World War I, even in cases where their personal lives had been shattered. Most returned to the institutions and laboratories where they had worked before 1914, and they took up the very same problems they had abandoned for war work in 1914 or 1915. Ernest Rutherford, Jean-Baptiste Perrin, Max Planck, and Gilbert Newton Lewis all are cases in point. The journal literature of 1920 differs very little from the literature of 1913, and articles and books written during the war years were widely read and first incorporated into research agendas in the early 1920s.

The development during the 1920s of the nonclassical tenets of quantum theory constituted a conceptual "revolution" in science, but not one that immediately undermined the conceptual authority of science, for either its practitioners or its public. The general theory of relativity, for all its bowdlerization by journalists, only reinforced respect for scientific authority, as demonstrated in the fascination with and mythologization of Albert Einstein.

After 1940 things changed more dramatically than after 1919. As a consequence, the years from 1800 to 1940 have the coherence for the history of chemistry and physics that is the basis for this volume's periodization. This history is mainly about science and scientists in Western Europe and North America from the time of Napoleon I and Queen Victoria to the outbreak of World War II. It focuses mainly on institutions in France, Germany, Great Britain, and the United States. The history of the successes and failures of internationalism, in conflict with powerful nationalisms, constitutes one of this book's important themes.

This book differs from other such histories in treating *both* modern chemistry *and* modern physics. A real effort has been made to show the similarities, differences, and connections between these two disciplines as they developed out of eighteenth-century traditions of natural philosophy, natural history, and chemical philosophy.

Some very fine histories of science that sweep chronologically from antiquity to the present include physics and chemistry within their purview, and there are excellent histories of physics and histories of chemistry that concentrate on the nineteenth and twentieth centuries (see the bibliographical essay).

This book aims to bring under a single compass many of the intellectual themes that are treated in these histories. It is neither a survey nor an exhaustive history, however, and much has been omitted in the interest of maintaining a manageable number of historical subjects. These include the development of a distinction between the chemical and physical atom; the conceptual competition between theories of the discrete corpuscle (or quantum) and theories of the continuum; the evolution of a field dynamics and an energy dynamics from a Newtonian force dynamics; the ongoing interplay in chemical explanation of traditions of natural history (biology) and natural philosophy (physics); the drama of turn-of-the-century discoveries of electrons, X rays, and radioactivity; and the truly revolutionary character of quantum mechanics, relativity theory, and nuclear science.

But if much in this history turns on ideas, these ideas and their empirical manifestations were the product not only of imagination, reasoning, and craft among women and men studying natural phenomena but also of educations, careers, and ambitions in particular settings of local and national politics, technologies, and ideologies. Some of this is analyzed as well. This book does pay some attention to the difficulties faced by women chemists and physicists in Europe and North America and by black scientists in the United States. The broader global story of the physical sciences as practiced outside the North Atlantic context by both Europeans and non-Europeans is told in other volumes. The reader is referred, for example, to Lewis Pyenson's *Civilizing Mission: Exact Sciences and French Overseas Expansion, 1830–1940,* the third in his series of volumes on science and colonialism. After 1940, it is impossible to ignore developments in chemistry and physics that took place in Japan, India, the Soviet Union, and other countries outside the North Atlantic sphere.

1

Disciplinary Organization in Nineteenth-Century Chemistry and Physics

Discoveries and controversies in physics and chemistry, the transmission and exchange of scientific information, and alliances, collaborations, and rivalries between and among physicists and chemists during the nineteenth century must all be set within their national and institutional contexts to be properly understood. This opening chapter, therefore, outlines the emergence and maturation of the institutions within which work in the physical and chemical sciences was performed during the nineteenth century, concentrating its attention on the development of university-level programs, research laboratories, and learned societies for physicists and chemists on the European continent, in Great Britain, and the United States. It briefly examines the development of periodicals devoted specifically to physics and chemistry and, finally, looks at the public role that these sciences had begun to assume by the nineteenth century's end. Discussion of the specific issues motivating research in physics and chemistry during the nineteenth century begins with chapter 2.

National Trends and Rivalries

In the early 1800s, institutional arrangements for the physical sciences were based in universities, government-sponsored engineering schools, academies and learned societies, and museums and observatories. Newer colleges, which oriented their curricula toward secular and practical education, proliferated in the second third of the nineteenth century, as did both specialized scientific societies and umbrella scientific organizations with national memberships. The weekly journal of the French Académie des Sciences (the *Comptes rendus hebdomadaires de l'Académie des Sciences*), which began to appear in 1835, set a new pace for a closely linked international network of communication among experimental and mathematical scientists.

During the course of the nineteenth and twentieth centuries, as the nation-state became the fundamental unit of political and military organization, nationalism and chauvinism played important roles in fostering scientific rivalries among scientists themselves and among their patrons and clients in government and industry. Scientific and technological achievements increasingly became indicators of national power and prestige. Napoleon Bonaparte fostered science and technology in France, as did Albert, the German-born prince consort of Queen Victoria, in Great Britain.

Scientists alleged a "decline of science" in Britain in the 1820s, comparing British science and mathematics unfavorably to French science, just as French scientists became concerned in the 1860s and 1870s about what they saw as the unfavorable condition of French science in comparison to that of the German states, which united under Otto von Bismarck and defeated France in military combat in 1870.

It is striking, however, that by 1900 the research productivity of physicists, in measured numbers of scientific papers, was approximately the same in England, France, Germany, the Netherlands, and the United States (and also in Japan). If allowance is made for the differences in these countries' populations, the numbers of physicists and physics students were also remarkably similar, as were expenditures for laboratories.

Most physicists were teaching or doing research in educational institutions, although some were entering engineering fields, especially as gas and electricity utilities burgeoned in the 1880s and 1890s. There was considerable consensus about the set of problems that defined the discipline of physics by 1900, including consensus on how to go about solving problems and on the best methods for training the next generation of practitioners.

In the field of chemistry, Germany by 1900 was widely acknowledged to be far ahead of other countries in the numbers of chemists trained in its institutions, the numbers of papers they produced, and the paid positions they held in industrial companies, including BASF (Badische und Soda-Fabrik), Höchst, Bayer, and AGFA (Aktiengesellschaft für Anilinfabrikation). These companies' success came principally from synthetic organic chemistry (based in the coal tar industry) for manufacture of dyestuffs and pharmaceuticals.

While the first steps toward the creation of industrial research laboratories were taken in the 1860s by William Henry Perkin's aniline dye company in Britain and by the La Fuchsine SA in France, it was BASF in the 1870s that first established clearly defined and long-lived "scientific laboratories" staffed by "theoretical chemists."

Early in World War I, the fact that the manufacturers of French military uniforms depended on German industry for their bright red dyes caused a scandal among the Allies. Even greater outrage erupted with the news that Fritz Haber, a Berliner who led German research in gas warfare, had received the Nobel Prize in 1918. His ammonia synthesis had been put into production at BASF during the war so that Germany, subject to the Allies' naval blockade, did not have to import nitrates for making explosives.

In the early decades of the nineteenth century, students of physics and chemistry, as well as of physiology, medicine, and mathematics, had come to Paris to study if it was at all possible for them to make the trip. They congregated at the lectures of the professors of the École Polytechnique, the Sorbonne, the Museum of Natural History, and the Collège de France, and they competed for permission to work in the private laboratories of Joseph-Louis Gay-Lussac (1778–1850) and Jean-Baptiste Dumas (1800–1884).

By mid-century, many American and British students were studying chemistry in Germany and taking doctoral degrees from German universities. From 1840 to World War I, nearly eight hundred British and Americans earned doctoral degrees in chemistry at one of the twenty German universities, more than half at the universities of Göttingen, Leipzig, Heidelberg, and Berlin. At least thirty-nine British chemists came under the influence of Justus von Liebig (1803–1873), Robert Bunsen (1811–1899), or Johannes Wislicenus (1835–1902). In contrast, hardly any British students studied in Paris after the death of Charles-Adolphe Wurtz (1817–1884).

Englishmen studying in Germany sometimes directed research, including the work of other Englishmen in German university laboratories. For example, F. S. Kipping's laboratory work in the Munich laboratory of Adolf von Baeyer (1835–1917) was directed by a more senior Englishman, "Privatdozent" William Henry Perkin, Jr. (1860–1929). Later during the 1920s, insiders joked that the official language spoken in Niels Bohr's theoretical physics institute in Copenhagen was "broken English," but English was already spoken by a majority of students in some German university chemical laboratories in the nineteenth century.

French worries about this state of affairs were so great that a French delegation was sent to the United States in the 1890s to investigate the reasons for the loss of French prestige in higher education. They were told that Americans preferred small German university towns to the bustle and anonymity of Paris. They also heard some criticism of the loose morals of Parisian night life. Perhaps more tellingly, the French system of higher education, they were informed, did not offer a doctoral degree which could be acquired quickly and straightforwardly by writing and defending a thesis, as was the case in Germany.

Decentralization of French university life away from Paris helped attract back to France some of those who had defected, and a second kind of doctoral degree, which could be obtained more simply than the more prestigious *doctorat d'État,* was established in 1896. Whereas in 1895 the University of Paris enrolled 80 percent of foreign students in the sciences, by 1929 the figure had declined to 25 percent, with the University of Toulouse enrolling 21 percent and the University of Grenoble 31 percent.

German influence waned a little in the early 1900s. Edward Frankland Armstrong (1878–1945) was less keen on the caliber of German scientific work than his father Henry Edward Armstrong (1848–1937) had been: the elder Armstrong regarded the chemical laboratory run by Hermann Kolbe (1818–1884) at Leipzig as the best in the world in 1865. In contrast, the younger Armstrong complained in 1900 that the Germans were not keeping up with foreign publications and that most of the good students in the Berlin laboratories, particularly that run by Emil Fischer (1852–1919), were foreigners or men from fields other than chemistry. The young Armstrong found Germans unimaginative in their doctoral research, simply pursuing their degrees by following orders for the preparation "of ethyl, butyl, propyl, hexyl, etc. derivatives of a known methyl acid."[1]

Nor was Edward Armstrong's unenthusiastic reaction to German training unique during this period. Over the next three decades there was to be much less Anglo-German exchange and interchange in chemical education, partly for reasons that helped bring on World

War I or that flowed from that war. There was also the perception that German chemistry no longer represented the cutting edge of physical chemistry or, especially, of chemical *theory.* In the summer of 1904, Harry Clary Jones of the Johns Hopkins University wrote from Germany to his colleague Edgar Fahs Smith at the University of Pennsylvania:

> It has been just ten years since I left Germany as a student [with Wilhelm Ostwald in Leipzig], and I can see very marked changes during that time. Ostwald has lost all interest in the experimental sciences, and is devoting all of his time and energy to philosophy and painting. . . . [J. H.] van't Hoff is working along quietly in Berlin in a few small rooms, has no students and wants none, and thus it goes with the physical chemists. The Germans seem to be keeping up their former interest and activity in organic chemistry, . . . but, all in all, Germany has lost much of her prestige in chemistry. . . . I am fully convinced that there is more good work being done in America in physics, physical chemistry, and inorganic chemistry than in Germany.[2]

On the other hand, the pursuit of the new quantum mechanics at Göttingen (at Max Born's institute), Munich (Arnold Sommerfeld), and Leipzig (Werner Heisenberg) in the 1920s revived the British-American connection to Germany. Americans wanted to learn quantum mechanics firsthand. In addition, even before the infamous "civil service" laws of 1933, which forced many Jewish and leftist scientists out of Germany, American institutions were becoming sufficiently important and sufficiently wealthy to attract a stream of visiting lecturers and some permanent residents from Western and Central Europe.

These scientists and mathematicians came not only to the handful of older private east coast schools that had nurtured science at the turn of the century—Harvard, MIT, Yale, Columbia, Johns Hopkins, Pennsylvania—but also to the University of Chicago, to the land-grant universities in Wisconsin, Minnesota, and Michigan, to the flourishing University of California at Berkeley, and to the California Institute of Technology (Caltech) at Pasadena.

The theoretical physicist Ludwig Boltzmann (1844–1906), on a third trip to the United States in 1905, was among those foreign scientists who kept diaries of their observations and feelings about California. Although there was much he liked, there was much, too, that

Boltzmann found unsettling, including the Berkeley community's prohibition of wine and beer. Determined not to be thwarted, Boltzmann made his way to Oakland and "succeeded in smuggling a right large battery of wine bottles back to Berkeley." The cost of medical attention at the local hospital also floored him: "This was the most expensive luxury of the whole trip and robbed me of a few mixed pleasures."[3] Some things have changed dramatically in Berkeley, others not so dramatically, since Boltzmann's day.

Universities and Laboratory Research Schools on the Continent

Peripatetic scholars were nothing new in the world of culture and learning. Tradition has it that Cambridge University was founded in the thirteenth century by a group of disgruntled Oxford students and clerics who transplanted themselves there. The medieval universities of Paris, Toulouse, Oxford, Padua, and Bologna were established by clerics who journeyed about the continent to fulfill their missions. The students and lecturers of the early universities were also travelers.

In the early nineteenth century, a surprising number of medieval traditions persisted in the universities. Professors at Oxford University were still expected to be ordained ministers of the Church of England, and a religious test for graduation from the University of Cambridge was abolished only in 1871. The core of the curriculum remained classical and philosophical in nature. Education was intended to cultivate the virtues of gentlemanly behavior for young men who had no need of a specialized profession. One important characteristic of these gentlemen was that they did not work with their hands. Another was that they were generalists.

It was not in England but in France and in Germany that the physical sciences were first specialized and professionalized. This specialization can be seen in schools and universities and, even earlier, in the founding of the Académie Royale des Sciences in Paris in 1666. By 1699 the academy had six sections: geometry, mechanics, astronomy, chemistry, botany, and anatomy. Antoine Lavoisier (1743–1794), as a member of the chemistry section, helped create a seventh section—for general physics—in 1785. By 1803, specialization had proceeded far enough that the academy reorganized its now-eleven sections into two classes, the "mathematical sciences"and "physical sciences," and provided each class with its own secretary. Mathematical science included physics; physical science included chemistry.

Twenty-two French universities were dissolved during the Revolutionary period and then reconstituted under Napoleon as branches of a "University of France." While German universities included philosophical faculties for most of the nineteenth century, French university centers distinguished letters from sciences faculties. Each sciences faculty was required to maintain four professors, one each for differential and integral calculus, rational mechanics and astronomy, physical sciences, and natural history.

As numbers of university students and faculty expanded over the course of the late nineteenth and early twentieth centuries, especially in the 1880s and 1890s, specializations within scientific fields also proliferated, so that by the turn of the century the most important French sciences faculties taught two or more branches of chemistry, which might include mineral chemistry, organic chemistry, physical chemistry, electrochemistry, and/or applied chemistry. Typically at least two professors belonging to the sciences faculty specialized in physics, one in general physics oriented toward experimental physics, and the second in rational mechanics or applied mathematics.

Unlike their counterparts at Oxford and Cambridge, French professors often sought large audiences—composed not only of university students, but also of local secondary students and the general public—by scheduling their lectures in the evenings or late afternoons. A great deal of emphasis was put on rhetoric, style, and elegance of presentation in lecture courses by the 1840s and 1850s. Students with the secondary-level baccaulaureate degree were awarded university degrees, or *licences,* by passing examinations, usually cramming for these exams from published versions of the courses they (may have) heard. Until the late 1860s, passing a national exam called the *agrégation* was a more important passkey to a teaching career in higher education than was writing a thesis.

Reforms affecting the entire educational system were enacted in France under a new republican government in the 1870s and 1880s. These measures led to expanded enrollments at the universities. Closed courses, which could not be attended by the general public, were instituted for university students and a doctoral thesis demonstrating original research, not just erudition, became an absolute requirement for appointment to any university faculty by the 1880s. By the early 1900s institutes for applied sciences were established at many universities, especially provincial ones: these institutes drew their teaching staff from the sciences faculties, further specializing and professionalizing the sciences and creating a clearer distinction between science and engineering.

The French adopted many of these reforms after studying the German system of education, in which universities put a great deal of emphasis on independent research and the writing of a thesis. Ironically, the German system was itself the result of reforms taken fifty years earlier in a francophobic reaction against French power and influence during the Revolutionary and Napoleonic period (1789–1815)—about the same time that "decline of science" concerns were being voiced in England.

Just after Napoleon's victory over the Germans and the Austrians at Wagram in 1809, the new University of Berlin opened to 156 students. It had close ties to the Berlin Academy of Science and was wholly supported by the Prussian state budget. Like other Prussian universities, the University of Berlin enrolled students from secondary schools (*Gymnasia*) who benefited from the first compulsory and universal system of primary education in the world. The German system produced a higher level of literacy and culture among a broader spectrum of its population than was true anywhere else in the West, and it produced scientific teachers and scientists of the second and third rank, not just an elite first rank.

While the universities emphasized traditions of classical learning, the higher technical schools (*technische Hochschulen*) taught applied sciences to students from nonclassical *Realschulen*. By the end of the century, the *technische Hochschulen* were permitted to grant the doctoral degree based on original research. Under cultural minister Karl Altenstein (1770–1840), research became the primary criterion for appointment to all Prussian universities by 1840. Unlike the French system, the Prussian system had no national examinations for university students. Students tended to enter a university to study with a particular professor, who was expected to teach materials from his own recent research. Thus specializations developed at different universities according to the interests of the teachers. Unlike the French system, the German system encouraged students to follow courses, seminars, and laboratory work at several different institutions.

Seminars and laboratory exercises made German universities unique for much of the nineteenth century. Seminars were first offered by philologists in the philosophical faculties to students intending to teach classical languages in the *Gymnasia*. By the 1820s and 1830s seminars provided future *Gymnasium* teachers in mathematics and the natural sciences the opportunity to learn firsthand, and in small groups, how to teach subject matter, handle basic instruments, and duplicate simple experiments. Unless privately arranged, these seminars were regulated by printed statutes, announced along with university lectures, and supplied with a few books, instruments, and stipends or prizes for the students.

Among the most famous early seminars in the physical sciences were those of J. S. C. Schweigger (1779–1857) at Halle, Heinrich Weber (1842–1913) at Göttingen, and Franz Neumann (1798–1895) and Heinrich Dove (1803–1879) at Königsberg. These seminars all emphasized experimental physics, techniques of precise measurement, and the use of mathematics to express constant relationships in nature. Between 1825 and 1870 half of all German universities established physics seminars, many of them becoming kernels for teaching and research laboratories for well-funded institutes.

The establishment of teaching laboratories initially appeared more problematic in Germany than in France. At the very beginning of the nineteenth century, *Physik* was taught in the German university philosophical faculty and *Chemie* in the medical faculty. As late as 1840, both the medical and philosophical faculties at Berlin rejected Justus von Liebig's claim that chemistry should be taught in the philosophical faculty, arguing against him that a science dependent on laboratory instruction had no place in a philosophical faculty. At this time the idea of a physics laboratory hardly existed; the word *laboratory* implied research in chemistry.

In 1863 the University of Tübingen became the first German university to establish a mathematics and natural philosophy faculty independent of the law, medicine, and philosophy faculties, thus breaking up the old pattern of teaching natural science (*Naturwissenschaft*) within the philosophical faculty (*Wissenschaft*) and providing greater freedom to establish new scientific specialties without fighting objections by classicists and humanists. By then, however, chemical laboratories were well established in universities throughout Germany, the most famous of which was Liebig's first laboratory, established at Giessen in the state of Hesse-Darmstadt in 1826.

Liebig's laboratory set a style that was later duplicated by his American students, although not by Parisian colleagues or the gentlemen of Oxford and Cambridge. Liebig lived in the building that housed the laboratory, and the students spent their entire day in its precincts. No strolls along the boulevard or afternoon cricket for them! Liebig's laboratory caretaker complained that he could not get the students to leave. Liebig relied on older students to act as instructors for the more junior men, while Liebig spent his time on his own research and writing or talking with his older students about their daily reports on their work.

This became the pattern of organization for modern laboratories in Germany. The style was imported into England, not through the old universities in Oxford and Cambridge but through the Royal College of Chemistry in London, whose first director in 1845 was Liebig's German student A. W. Hofmann (1818–1892), who trained English chemists for twenty years before returning to Germany. Other of Liebig's students established important laboratories thoughout Europe: Friedrich Wöhler (1800–1882) in Göttingen, A. Strecker (1822–1871) in Oslo, Auguste Kekulé (Friedrich August Kekule von Stradonitz, 1829–1896) in Ghent and Bonn, Hermann Kopp (1817–1892) in Giessen and Heidelberg, Charles-Adolphe Wurtz in Paris, Lyon Playfair (1818–1898) in London and Edinburgh, Alexander Williamson (1824–1904) in London, and Edward Frankland (1825–1899) in Manchester and London.

The sociologist Jack Morrell has described Liebig's "school" at Giessen as a distinct break with past forms of organization of teaching and research in science. Liebig's laboratory was the first instance of a "knowledge factory."[4] The laboratory or institute director assigned topics for study, established techniques and instrumentation for the laboratory, arranged ready access to a medium of publication, and placed graduates in positions in universities, industry, and government. The exercise of authority, including control of the dissemination of research results, was a significant imperative for the master of a research school. Liebig's student Hermann Kolbe learned the lessons well. On taking up the editorship of the *Journal für praktische Chemie,* Kolbe wrote to a friend:

> What particularly attracts me about editing a journal is the value it has for a large chemical laboratory to have a journal at its disposal at all times . . . when it is a question of rapid publication and defense and advocacy of a viewpoint. . . . [I]t would be very unfortunate were it to end up in the hands of another director of a large chemical institute, such as Hofmann.[5]

At Liebig's laboratory, as at other research schools, there was a characteristic program and methodology for research. In Liebig's case, it was the isolation and characterization of organic substances and the application of chemistry to nutrition and agriculture. His students learned simple and easily duplicable techniques for quantitatively establishing the composition of substances. Results from the laboratory were published in the journal of pharmacy and chemistry *Liebigs Annalen* which Liebig launched in 1832 and which is published still today.

Physics and chemistry institutes proliferated in the late nineteenth century in Germany, evolving out of a combination of the seminar, the laboratory, and the colloquium. At Berlin, for example, Heinrich Gustav Magnus (1802–1870) coordinated his instructional seminar with his formal lectures. Because he found university rooms inadequate, he brought university apparatus to his house, where he combined it with his own apparatus and library; in 1863, he began designating this setup as the official Berlin "physical laboratory" for research. As early as 1843, he had hosted a weekly "physical colloquium" in his home for discussion and criticism by students of recent literature in physics.

In 1867 Magnus proposed that the university construct a proper physics institute, and, when Hermann von Helmholtz (1821–1894) succeeded Magnus as professor of physics at Berlin in 1870, he received assurances from the Prussian ministry that a new physics institute would be built for him. It was equipped with everything needed for teaching, his own research, and student exercises, including an auditorium and the director's living quarters.

In 1884 Helmholtz agreed to become director of the new Physikalisch Technische Reichsanstalt (PTR), to be built on the outskirts of Berlin in the bourgeois suburb of Charlottenburg. He worked with the institute's principal founder, Werner Siemens (a wealthy industrialist who was Helmholtz's father-in-law) and with Prussian state architects to design the physical plant. Begun in 1887, it consisted of ten buildings, five for the scientific section and five for the technical section. Including expenses for its magnetic house and machine house, the bill for the scientific section was almost 1 million marks, just a little less than the Prussian state's expenses for physical institutes at Berlin or Leipzig completed in the 1870s. An important difference was that the PTR's scientific section was targeted entirely for research, whereas most institutes in physics, theoretical physics, physical chemistry, and other disciplines were attached to universities and in principle served primarily pedagogical purposes.

The PTR became a model for later independent, state-funded research laboratories, including the National Physical Laboratory in England (est. 1900) and the National Bureau of Standards in the United States (1901). France, as in so many matters, did not adopt the German model but instead established an umbrella organization for funding researchers and their laboratories at traditional centers of higher education throughout France, the Centre Nationale des Recherches Scientifiques (1939).

Scientific Education and Careers in Great Britain and the United States

The German way of doing science gained renewed respect from scientists, entrepreneurs, and politicians in Great Britain following the Prussians' defeat of the Austrians at Sadowa in 1866 and their victory in the Franco-Prussian War in 1871. The Germans displayed increasing industrial and technological strengths at international expositions beginning with the Crystal Palace Exhibition in London in 1851 and, more spectacularly, at the Paris Exhibition of 1867. After a government-sponsored mission abroad to study continental secondary and higher education, Matthew Arnold (1822–1888) returned to England a Germanophile, convinced of the superiority of German education and culture:

> What I admire in Germany is, that while there, too, Industrialism, that great modern power, is making at Berlin and Leipzig and Elberfeld most successful and rapid progress, the idea of Culture, Culture of the true sort, is in Germany a living power also. Petty towns have a university whose teaching is famous through Europe; and the King of Prussia and Count Bismarck resist the loss of a great *savant* from Prussia as they would resist a political check.[6]

In Arnold's view, the universities of England, by which he meant Oxford and Cambridge, had no science at all.

Of course this was not perfectly true, but important research in physics and chemistry in Great Britain was taking place in Scotland and in English city-centers, not at Oxford or Cambridge. The cliché was fairly accurate until late in the nineteenth century that much of British science was done by amateurs—meaning men and women who did not have specialized educational credentials like members of the physical and mathematical sections of the Paris and Berlin academies of science.

At its founding, the Royal Society of London had no specialized disciplinary sections. Its 1662 letters of patent established the Royal Society to support philosophical studies, "especially those which endeavor by solid Experiments either to reform or improve Philosophy."[7] It was only in 1838 that the Royal Society established committees for special branches of science. The permanent committees included astronomy, chemistry, geology and mineralogy, mathematics, physics, and physiology, including the "natural history of organized beings." Electricity fell under the purview of the chemistry committee, which was initially chaired by Michael Faraday (1791–1867). Mechanics was addressed by the mathematics committee, and the physics committee concerned itself largely with meteorology and the "physics of the earth" in a tradition that has been called "Humboldtian science" in honor of the German naturalist and explorer Alexander von Humboldt (1769–1859).

These specializations were largely a product of the decline-of-science movement and the organization in 1831 of a new national subscription society, the British Association for the Advancement of Science (BAAS). It was modeled on the Deutsche Naturforscher-Versammlung, which had been meeting annually in different German cities since 1822, and which organized its members in scientific sections. Shortly after the founding of the BAAS, William Whewell (1794–1866) coined the English word "scientist" to describe who it was who attended the meetings.

The old universities at Oxford and Cambridge began accommodating themselves to demands for scientific education in the 1830s. By the late 1840s Whewell had changed the subject content of the final university examinations at Cambridge, called the "mathematical tripos," to include greater emphasis on "mixed mathematics," namely, mechanics, hydrodynamics, planetary theory, and optics. When he was a tripos examiner in the late 1860s, James Clerk Maxwell (1831–1879) broadened the problems on the exam to include current subjects in heat, electricity, and magnetism. Nearly half the chairs in physics in British universities in the second half of the nineteenth century were held by "wranglers," men who had obtained first-class degrees by ranking at the top of the Cambridge mathematical tripos. These included Englishmen such as George Gabriel Stokes (1819–1903) and Joseph Larmor (1857–1942) and

Scotsmen such as William Thomson (later Lord Kelvin, 1824–1907) and Maxwell.

The natural sciences tripos, emphasizing experimental physics, chemistry, and natural history, was established at Cambridge in 1851, and an honors school in the natural sciences was founded at Oxford about the same time. As in Germany, France, and the United States, instructional and research laboratories began to be built on a palatial scale in England in the 1870s, including the Clarendon Laboratory at Oxford and the Cavendish Laboratory at Cambridge, modeled partly on Lord Kelvin's laboratory at Glasgow.

Both Glasgow and Edinburgh, as well as Dublin, had strong traditions in the physical sciences. As early as 1829 Thomas Thomson (1773–1852) began practical chemical training in a university-sponsored chemical laboratory at Glasgow. Influenced by his experiences in the Parisian laboratory of Henri-Victor Regnault (1810–1878), William Thomson in 1850 organized a physical laboratory near his lecture room at Glasgow, which remained the only academic site for theoretical and practical instruction in electricity in Great Britain until the 1860s, when George Carey Foster (1835–1919) established a laboratory at University College, London. Much of the work at the Cavendish Laboratory under Maxwell and his immediate successor John William Strutt (Lord Rayleigh, 1842–1919) followed Kelvin's lead, concentrating on precise measurement of physical constants, using electrical devices and spectroscopes.

The number of academic physics laboratories grew in Britain from ten in 1874 to twenty-four by 1885, a period in which the setting up of applied-science professorships and laboratories was differentiating physics from engineering in Great Britain. Chemical laboratories also proliferated, most of them, like the physics laboratories, established at the so-called new colleges or redbrick universities of London and the industrial Midlands.

At Manchester, where John Dalton (1766–1844) had struggled to find students to tutor in the early 1800s, the new physics laboratory completed in the early 1900s was the fourth-largest physics laboratory in the world after those at the Johns Hopkins University in Baltimore and the universities at Darmstadt and at German-occupied Strassburg. The chemists at Manchester had enjoyed good buildings since 1872, when Henry Roscoe (1833–1915), a student of Thomas Graham (1805–1869) and Alexander Williamson in London and of Robert Bunsen in Heidelberg, designed a facility that included his own private laboratory, two teaching laboratories, and more than twenty small research laboratories and staff rooms. By 1887 the number of students in the department had reached 120, and the department had an honors program for the best of them. One of Roscoe's assistants, Carl Schorlemmer (1834–1892), a student of Heinrich Will (1812–1890) and Hermann Kopp, became the first chairholder in organic chemistry in Great Britain, at Manchester, in 1874. By 1900 the Schorlemmer Laboratory and the Perkin Laboratory (named for William H. Perkin) provided enviable facilities for organic teaching and research.

Dissenters (Protestants who had left the Church of England) and the industrial and agricultural entrepreneurs of London and the Midlands were responsible for the founding of the universities outside Oxbridge, and it was they who paid many of the expenses for academic buildings and scientific laboratories. Utilitarians led by philosopher Jeremy Bentham (1748–1832) and jurist Henry Brougham (1778–1868), opened University College in London in 1828, with open admissions to all men, including Dissenters, Catholics, and Jews. Women were admitted in 1878, and William Henry Bragg's laboratory at the college was notable in the 1920s not only for its advances in X-ray crystallography but also for its inclusion of women researchers.

King's College was set up in London as an Anglican educational alternative in the metropolis. Professors, although not students, had to be members of the Church of England. They included the geologist Charles Lyell (1797–1875) and the physicist Charles Wheatstone (1802–1875). The University of London Act of 1898 established a structure for the many institutions that had been founded over the course of the previous century, including Bedford College and Westfield College for women, the medical schools of the principal London hospitals, the London School of Economics and Political Science, and the Imperial College of Science and Technology.

Imperial College, built in the suburb of South Kensington in the 1880s and connected to central London by the first subway line in the world, was a conglomeration of scientific and technical institutions: the Royal College of Chemistry (est. 1845), the Royal School of Mines (1851), the Finsbury Technical College (1883), Central Technical College (1884), and others. Like its counterparts in Manchester, Liverpool, Leeds, Birmingham, and Sheffield, Imperial College grew out of a piecemeal array of institutions, with shorter or longer lives, that provided demonstration-lectures for the genteel public, mechanics institutes for the working classes, and small colleges for the children of middle-class Dissenters.

Owens College at Manchester, for example, owed its origin in 1851 to John Owens's trust fund, which provided monies for an institution to instruct "young persons of the male sex" in "such branches of learning and science as are or may be taught in the English Universities"; like University College in London, it was free from religious tests.[8] Unlike the young men who matriculated at Oxford and Cambridge, those who enrolled in Manchester often had strong interest in science and its industrial applications. The University of Manchester maintained an unbroken line of eminent chemists, including Frankland, Roscoe, Schorlemmer, W. H. Perkin, Jr., Arthur Lapworth (1872–1941), Chaim Weizmann (1874–1952), Robert Robinson (1886–1975), and, in the early 1930s, Michael Polanyi (1891–1976). Linus Pauling and Robert Millikan rightly despaired in 1938 of enticing Alexander Todd to Caltech when they heard he was to be offered an opening at Manchester, which they knew to be "much more in the chemical swim than Caltech."[9]

The English redbrick universities were also the institutions most open to women who wanted to pursue scientific education and careers. In the United States, women's colleges such as Vassar (est. 1865), Radcliffe (1879), and Bryn Mawr (1885) provided excellent opportunities for scientific education and careers. Women's colleges began to be established at Cambridge in 1869 and at Oxford in 1879, but the University of Oxford did not admit women to full membership until 1918—and the University of Cambridge not until 1948.

The American geologist Florence Bascom was one of the first women to receive a doctoral degree, taking a Ph.D. at Johns Hopkins in 1893. She taught geology at Bryn Mawr, where Ida Hyde was an undergraduate before becoming one of the first women to receive a Ph.D. in Germany. The University of Chicago, founded in 1891 as a private institution for men and women with a strong emphasis on graduate instruction and research, provided a coeducational environment in the United States, as did some of the land-grant institutions.

Many women found it easier to pursue research for a doctoral thesis in foreign institutions than in their own countries, so that women from the United States and Central Europe were frequently to be found in German and French laboratories, where their teachers expected them to return home following the degree, not to set up shop in the adopted country. Marie Curie (née Manya Skłodowska, 1867–1934), who came to Paris from Poland to study mathematical and experimental physics under Gabriel Lippmann (1845–1921), is the best-known example of a woman who did the unexpected. Marie Curie succeeded her French husband, Pierre Curie (1859–1906) in his physics professorship at the Sorbonne but never enjoyed election into the Academy of Sciences.

Many women collaborated with their husbands, whose institutional influence could be used to acquire laboratory workspace for the nonsalaried wife. This was true, for example, for the chemists Gertrude Walsh Robinson (1891–1962), the wife of Robert Robinson (1886–1975), and Edith Hilda Ingold (1898–1988), the wife of Christopher K. Ingold (1893–1970). Still, well into the twentieth century, the prevailing attitude persisted that women with careers should not be married and women who were married should remain amateurs at science. This prevailing attitude can be seen in the remarks of Henry Edward Armstrong in his presidential address to the BAAS in Winnipeg in 1909:

> If there be any truth in the doctrine of hereditary genius, the very women who have shown ability as chemists should be withdrawn from the temptation to become absorbed in the work, for fear of sacrificing their womanhood; they are those who should be regarded as chosen people, as destined to be the mothers of future chemists of ability.[10]

The career of one of Armstrong's own students, Ida Smedley, [later Maclean] (1877–1944), demonstrates one of the career patterns for women scientists in the early 1900s. After studying in Armstrong's laboratory at the Central Technical College, she became an associate at Newnham College in Cambridge from 1899 to 1903, a researcher at the Davy-Faraday Research Laboratory in 1905, an assistant lecturer in chemistry at Manchester in 1906, and then a member of the research staff in the biochemical department at the Lister Institute. In 1915, she received the Ellen Richards Research Prize, awarded by the Naples Table Association, for the best scientific thesis written by a woman. Her research was incorporated into a book coauthored with her husband, Hugh MacLean, *The Lecithins and Allied Substances* (1918). Thus, her career followed the example of "collaborative couples."

A different but equally typical example was that of Jane M. Dewey, the youngest daughter of the American philosopher John Dewey, who completed a doctoral thesis in physical chemistry at MIT. Following two years at Niels Bohr's institute for theoretical physics in

Copenhagen, she was awarded a National Research Council fellowship to Princeton University, where the physicists referred to her as "Magie's folly." (Dean William F. Magie had argued in favor of her coming to Princeton.) The letters of recommendation from her MIT mentor, Karl Compton, resulted in no job offers, with one physicist at Berkeley reporting that his colleagues would not permit a woman in the department. Eventually Dewey took a position at Bryn Mawr.

Dewey at least had a paid faculty post, in contrast to many women who received no stipend or only negligible remuneration for their scientific work. Historian Margaret Rossiter has found that in the United States, the number of women holding titles such as "research associate" increased from the 45, as listed in the 1921 volume of *American Men of Science* (16 percent of the total listings) to 282 (22 percent of the total) in 1938. Robert Millikan, head of Caltech from 1921 to 1945, was of the opinion that no good job should be "wasted" on a woman.

Like women, Jewish scientists met discriminatory practices in the United States, Western Europe, and Great Britain. In 1921 Millikan hired Arnold Sommerfeld's student Paul S. Epstein (1883–1966) "even though [he] is a Jew" while declining to hire another Jewish physicist because Millikan thought it better not to have "more than about one Jew anyway."[11] In 1930 an officer at the Rockefeller Foundation expressed the same reservation about giving postdoctoral fellowships to Jews as to women: that not many should be granted since these scientists had such difficulty in getting positions later. These kinds of objections to Jewish scientists largely disappeared after World War II, both because of the many well-known contributions made by Jewish scientists to the Manhattan Project and because of the realization of the horrible consequences of Nazi anti-Semitism.

Colleges established as Negro institutions provided education, research facilities, and jobs for some black American scientists. By 1940 American black colleges numbered about fifty-five, mostly in the South, with Howard, Morehouse, Fisk, and Tuskegee among the most prominent institutions.

In 1863 Yale University conferred the first American Ph.D. in the sciences (actually, in engineering) on Josiah Willard Gibbs. And in 1876 Yale became the first American university to grant a doctoral degree to a black American, Edward Alexander Bouchet (1852–1918), who received the degree in physics and who went on to teach physics and chemistry in high schools. Following Bouchet, approximately 125 black Americans took Ph.D.s in the sciences from 1876 to

1943; 96 percent of these scientists would at some point in their careers teach at a Negro college or university. Among black American scientists during this period, about a dozen completed doctorates in physics and perhaps four times that number in chemistry.

One of the most successful black American chemists was Percy Julian (1899–1975), who took his Ph.D. in organic chemistry in Vienna in 1931 and taught at Howard University and then DePauw University before becoming director of research at the Glidden Company from 1936 to 1945. In 1953 he established his own fine-chemicals company, Julian Laboratories in Illinois. Julian and his DePauw colleague Josef Pikl carried out the first total synthesis of the indole alkaloid physostigmine in 1935 at the time Robert Robinson and his coworkers at Oxford were attempting to perform this same kind of synthesis.

By the early twentieth century it was beginning to be recognized in Britain and Europe as well as in non-European, scientifically oriented countries such as Japan, that the United States was developing a scientific and engineering infrastructure that rivaled European centers. Many Americans aspiring to careers in chemistry and physics were now completing their master's and doctoral work at home. Indeed during 1910–1911 the Prussian education ministry compared twelve American public and private universities with the twenty-one universities in the German system and found that the average budget of a German university was 1.67 million marks, whereas the average for the twelve American schools was more than three times greater, the equivalent of 5.80 million marks annually.

Europeans also noted with envy the role played in American scientific research by private foundations, especially the Rockefeller Institute for Medical Research in New York City and the Carnegie Institution in Washington, D.C., both of which supported science at home and abroad. During the period from 1902 to 1916, the Carnegie Institution provided $218,000 in funds for chemical research, 85 percent of which went to physical chemistry. The Rockefeller Foundation (established 1919), along with the National Research Council, provided postdoctoral support to 20 percent of all American physics Ph.D.s and 5 to 10 percent of all chemistry Ph.D.s during the 1920s, most of whom now chose to take their doctoral work at home before doing a year or two of postdoctoral study abroad.

In 1914, Theodore William Richards (1868–1928) of Harvard University became the first American to receive a Nobel Prize in Chemistry, honored for his precise determination of atomic weights. Even before this, in 1901, Richards had been granted an unprecedented distinction among American chemists by being offered the chair of physical chemistry at the University of Göttingen. Duplicating Liebig's old ploy of using an outside offer to better his position, Richards obtained a promotion from Harvard, a reduction in his teaching load (to three lectures a week), and three-quarters of a million dollars for what became the Wolcott Gibbs Memorial Laboratory, said to be among the best equipped chemical laboratories in the world at that time.

American research schools of considerable merit were emerging not only in the older private universities of the East but also at midwestern and west-coast institutions. At the University of Illinois, Harvard-educated Roger Adams (1889–1971) trained more than two hundred Ph.D. and postdoctoral students in chemistry, some of whom, like Wallace H. Carothers at Du Pont, were to make spectacular careers in industry,

By the time the Gibbs Laboratory opened at Harvard in 1912, both Harvard and MIT were having to fend off raids from science departments on the west coast that were poised to become major research schools in the physical sciences in the 1920s. The young Gilbert Newton Lewis (1875–1946), for example, began attracting to his new Berkeley laboratory some of the instructors and assistants he had known at MIT: William C. Bray, Richard Chace Tolman, Joel Hildebrand, and Merle Randall, among others.

Under Lewis's leadership from 1912 to 1937, the Berkeley chemistry school strongly emphasized physical chemistry, producing 290 Ph.D.s and 5 Nobel laureates.[12] Indeed, by the mid-1920s, physical chemists like Lewis were beginning to challenge organic chemists for leadership in the American chemical community. More than 25 percent of the papers that appeared in the *Journal of the American Chemical Society* were in the area of physical chemistry. Well before what many consider the turning point represented by World War II, the United States had become a major scientific power.

The Development of Expertise: Scientific Societies and Journals

The Chemical Society of London, over which Armstrong exerted so much influence for so many decades, and the American Chemical Society, over which G. N. Lewis was to have similar authority, were among the many specialized scientific societies established in the mid- and late nineteenth century.

Previously, the Royal Society of London, the Manchester Literary and Philosophical Society, the American Philosophical Society, the Berlin Academy of Sciences, and other learned societies had provided meeting grounds for elected members interesting in discussing a broad range of topics in mathematics, the natural and physical sciences, and natural philosophy. The weekly meetings of all the members of the Paris Academy of Sciences, for example, ensured that specialists in one field remained abreast of developments in another.

In addition, national subscription societies such as the BAAS and the American Association for the Advancement of Science, founded in 1848, provided forums that moved from city to city for annual meetings or conventions. These were open to scientists at all ranks of expertise, independent of institutional affiliations and credentials.

But the desire of scientists to meet as smaller groups of specialists was already being registered in the early 1800s in the founding of organizations such as the Linnaean Society (for the study of natural history, est. 1788), the Geological Society of London (1807), and some other new groups.

The first of the long-lived specialized societies in the physical and chemical sciences was the Chemical Society of London, founded in 1841. Other long-lived chemical societies were established in the decades following the founding of the London Chemical Society, among them the Société Chimique de Paris in 1857, the Deutsche Chemische Gesellschaft in 1867, and the American Chemical Society in 1876. These societies all established journals for publishing current research in the field. The London Chemical Society's journal (1847) was at first a quarterly, but by 1861 the need for a monthly journal had become clear. These chemical journals more and more required precise accounts of experimental results, refused papers of a purely theoretical nature, and inundated their readers with papers in the field of organic chemistry.

Not all the journals were well received. The early *Journal of the American Chemical Society* catered mainly to New York City members and was not highly regarded. Until a reform of the American Chemical Society in the 1890s, the *American Chemical Journal,* edited independently at Johns Hopkins by Wöhler's student Ira Remsen (1896–1927), was instead the most highly regarded American chemical publication.

Like Remsen's journal, many of the most influential journals in the physical sciences were not tied to a membership society, but were edited by entrepreneurial scientists, usually ones with laboratories. Among the most important chemical journals internationally were *Liebigs Annalen,* already mentioned, and the *Annales de chimie et de physique.* Like the German *Annalen,* the French *Annales* has had a long history. Founded by Lavoisier as a journal for chemistry, it was continued as a journal for chemistry and physics by Gay-Lussac and François Arago (1786–1853), and in 1914 it split into the *Annales de Chimie* and the *Annales de Physique.*

In addition to the *Annales de Physique,* the American journal *The Physical Review,* a private journal that was adopted by the American Physical Society in 1913, was also beginning its rise to prominence. Another influential physics journal had been launched by F. A. C. Gren in 1790 under the name *Annalen der Physik und Chemie.* L. W. Gilbert eliminated "Chemie" from the journal's name, then restored it in the form of "Physikalischen Chemie" during the years 1819 to 1823, before J. C. Poggendorff restored "Chemie" without qualification in 1824, on the grounds that physics and chemistry could not be separated. Many highly mathematical articles by G. S. Ohm, Wilhelm Weber, Franz Neumann, and Thomas Seebeck filled its pages. Following the founding of new physical chemistry journals in the 1880s and 1890s, the *Annalen* again dropped "Chemie" from its title in 1900.

Like chemical societies, physics societies began to proliferate at mid-century. The first physics (or "physical") society was the Deutsche Physikalische Gesellschaft, founded in 1845. The society's concerns initially included general physics, acoustics, optics, electricity, heat, and "practical physics" meaning meteorology and earth science. The Société Française de Physique was established in 1873. And the Physical Society of London was founded the next year by physics-oriented members of the convivial "B Club," named after Section B (Chemistry) of the BAAS.

The Physical Sciences and Public Needs

Besides scientific academies and societies, technical colleges and university institutions, and industrial laboratories, government-sponsored bureaus remained important sources of scientific research and scientific careers throughout the nineteenth and early twentieth centuries, both in military and civilian affairs. Geological surveys, agricultural experimental stations, bureaus of weights and measures, and patent offices were among the kinds of government offices that, sometimes unknowingly or reluctantly, made physical and chemical research possible. One well-known case of this is the work accomplished in theoretical physics by Albert Einstein (1879–1955) while he earned his living by deciding the fate of patent applications for electrical devices in a Bern patent office.

A less well known but more typical example is the organization in 1900 of a laboratory of chemistry and physics at the U.S. Geological Survey, staffed by men trained in the United States and Europe. One of these scientists, Arthur L. Day (1869–1960), had worked on high-temperature thermometry at the Physikalisch Technische Reichsanstalt in Charlottenburg, and another USGS staff member had worked at Walther Nernst's Institute of Physical Chemistry in Göttingen. Under Day's leadership, the USGS laboratory staff carried out studies of silicate systems under controlled laboratory conditions. The result was the new field of geophysics.

Museums represented another important type of government institution that often provided opportunities for research. Usually these were natural history museums, such as the Muséum d'Histoire Naturelle in Paris or the Deutsches Museum in Munich. The holder of the chair of physics at the Paris museum gave well-attended public lectures and carried on whatever research suited him. Four generations of the Becquerel family held this position; the third, Henri Becquerel (1852–1908), shared the 1903 Nobel Prize in physics with the Curies for the discovery of radioactivity. His son, Jean Becquerel (1878–1953), had a distinguished career that included work in low-temperature physics.

In the United States, the *Reports* issued by the Smithsonian Institution in Washington, D.C., became an important source of summaries of the latest physical research around the world at the turn of the century. The Smithsonian was an institution that nearly refused to accept its birthright. James Smithson (1765–1829), the illegitimate son of the Duke of Northumberland, was a fellow of the Royal Society, and a chemist and mineralogist. His will, probated in 1835, directed that in case of the death of his nephew, which occurred, his one-half-million-dollar estate was to be bequeathed to the U.S. government for the founding of an institution to further science. Senator John Calhoun and others opposed the American government's accepting the money, while Representative John Quincy Adams, who had been President of the United States from 1825 to 1829, ultimately persuaded Congress to do so. Legislation establishing the Smithsonian Institution was finally enacted in 1846.

Just as it was an Englishman who was responsible for founding the Smithsonian Institution, it was an American who founded one of the great British scientific institutions, the Royal Institution in London. Benjamin Thompson (1753–1814) was a Loyalist who left New England during the Revolutionary War. Knighted by the Elector of Bavaria in 1784 for his services to the Bavarian state, it was as Count von Rumford that he persuaded London friends that the English capital needed an institution comparable to the Conservatoire Nationale des Arts et Métiers, established in 1794 as a Sorbonne for working men.

The Royal Institution received a charter from George III in 1800 but received no more funding than the Royal Society of London had gotten from Charles II in 1665. The aims of the institution were twofold: "the general diffusion of the knowledge of all new improvements, in whatever quarter of the world they may originate; and teaching the application of scientific discoveries to the improvement of arts and manufactures in this country, and to the increase of domestic comfort and convenience.[13] Humphry Davy (1778–1829), who succeeded Thomas Garrett as professor of chemistry at the Royal Institution, set the institution on its successful future path of public lecture courses and private scientific investigations, attracting large numbers of annual subscribers and generous gifts from wealthy patrons, who attended lectures, visited the library, and occasionally requested advice.

The Royal Institution was governed by a board of managers that gradually became more professionally knowledgable about science. With time, endowments were established for the Institution, so that by 1833 the Fuller endowment provided the professor of chemistry, Michael Faraday, with an endowed salary. Faraday had laboratory facilities, apparatus, materials, and an assistant at his disposal. He and his wife were provided with residential rooms, including coals and candles. Faraday and the other professors offered an array of lectures: morning laboratory lectures to London medical students, an afternoon course of six to eight lectures during the winter and spring, the Friday Evening Discourses (about four each season), and the Christmas series of lectures for young people.

Faraday and others often placed themselves at the disposal of their patrons and of government and commercial interests, advising, consulting, analyzing, and experimenting. For example, Faraday was called on for advice following an explosion in 1843 at the Royal Gunpowder Factory at Waltham Abbey, and he was appointed by the British home secretary in 1844 to investigate the Haswell Colliery explosion. He resisted efforts to serve on committees concerned with the ventilation of the new Houses of Parliament, but in many respects he demonstrated the usefulness of scientific knowledge for the development and management of a modernizing society. While some historians have questioned whether Faraday and other lecturers at the Royal Institution spent too much time at the service of public powers and entreprenurial interests, the Royal Institution can be seen as a successful venture in balancing what came to be called pure and applied science for the benefit of the larger public.

The Age of Positivist Science and the Cultural Symbolism of the Nobel Prize

The nineteenth century has sometimes been called the Age of Science. Such striking reforms took place within education that both the organization of educational institutions and the subject matter taught changed more in the course of one century than they had from the medieval period to the outbreak of the French Revolution. The changes that took place from the early to mid-twentieth century were not nearly so radical; rather, they merely continued trends that were well in place by the late 1870s.

All the physical sciences were affected by common, characteristic trends: the invention of laboratory instruction, the emphasis on precision and quantification in experiment, the mastery of a well-defined field of special knowledge, the desire for new results and their rapid communication, and the concern for public approbation and support of scientific activity. Increasing specialization and professionalization within science (*Wissenschaft*) dovetailed with the creation of new kinds of institutions and the carving up of subject matter within narrower and narrower disciplinary and subdisciplinary boundaries.

At the end of the eighteenth century, in reaction against what some saw as the excesses of Enlightenment rationalism, a Romantic reaction against the reduction of the natural world to mechanisms and mathematics had powerfully affected literature, philosophy, and the arts. This Romantic reaction was longer-lived in Germany than elsewhere, and many German scientists teaching at mid-century viewed their youth as a period of struggle against what Liebig dramatically called "the black death of our century," by which he meant Romantic *Naturphilosophie*.

Similarly, Helmholtz blamed the backwardness of German medicine in the 1840s on the quackery of Romantic theories about polarities, equilibrium, and holism. Even Johann Wolfgang von Goethe (1749–1832), whose literary masterpieces and scientific writings came to epitomize the Romantic movement, criticized *Naturphilosophie* at the end of his life.

The nineteenth century was to belong not to the romantics, but to rationalists, the greatest of whom was the Frenchman Auguste Comte (1798–1854), who was educated at the École Polytechnique in the early 1800s and who became absolutely convinced of the virtues of the scientific age, which he called "positive." The nineteenth century was to be the "positivist age."

For Comte, all cultures evolve through necessary stages of mental growth, from the theological (gods), to the metaphysical (forces), to the positive (facts). French culture and Western scientific culture, he thought, had completed the first two stages of historical development and were entering the third and final era. This positive stage would require a division of labor among scientific specialties, with broad categories corresponding to the fields of mathematics, physics, chemistry, biology, and sociology, which had emerged in that historical order. Comte's influence was great because it reflected the era's assumptions about the methodology and value of the sciences, as well as about the role of science in secularizing the social order.

The subsequent history of scientific education and research in the nineteenth century, with disciplinary specialization and proliferation of scientific institutions, seemed to follow the lines that Comte had predicted. So, too, physical scientists frequently claimed to work according to a Comtian methodology that emphasized facts, quantification, and analogies—or what came to be called positivist empiricism—in contrast to a methodology of hypothesis that multiplied invisible entities. Drawn from the Laplacian program of physics (see chapter 2), Comte's positivism was a program of unification of scientific explanation based on mathematical and physical principles.

By the early twentieth century, some one thousand physicists and three thousand chemists were active in Europe and North America, with that number increasing to three or four thousand physicists and three or four times that many chemists by the mid-1930s. As the activities of these men and women became more specialized, more professionalized, and less academic, rewards and recognition became proportionally more important both for the scientists themselves and for the institutions that employed them or invited them into membership.

Nothing in the twentieth century came to epitomize more thoroughly the prestige and the achievement of the sciences than the annual Nobel Prizes established through the will of the Swedish chemist, engineer, and industrialist Alfred B. Nobel (1833–1896). The prizes were to be awarded "to those who, during the preceding year, shall have conferred the greatest benefit on mankind." Members of the Royal Swedish Academy of Sciences were designated to decide awards for physics and for chemistry; the Royal Caroline Medico-Chirurgical Institute for physiology or medicine; the Swedish Academy for literature; and the Norwegian Nobel Committee (appointed by the Norwegian parliament) for peace.

The result was not only new international prestige for the Swedish scientific and literary communities, but also an extraordinary new role for members of the Swedish Academy of Sciences in shaping the definition and the history of what came to be counted as significant scientific work. For example, three of the first chemistry awards from 1901 through 1904 went to chemists identified with experimental work in physical chemistry, helping to establish the legitimacy of what was still a controversial field among many organic chemists. The Nobel awards' emphasis on experimental discoveries and investigations and on the establishment of physical and chemical *laws* rather than theories reinforced the empiricist and positivist heritage of the nineteenth century, a tradition that was already strong among Swedish physicists and chemists.

The symbol of the Nobel Prize became an important means of legitimation and recognition for scientific work, identifying science with social progress and with individual genius. From the early nineteenth century onward the history of chemistry and physics is a record of steady growth in the numbers of men and women educated and employed in scientific fields as well as in the numbers of discoveries and inventions they achieved. With increasing numbers of scientists and increasing specializations, a breakdown in the common ethos of the sciences might be expected. But, the Nobel language of "benefit to mankind" reiterated the claims of scientists to discover and create knowledge that is valuable for both its beauty and its usefulness. The Nobel awards' rituals of selection reproduced the social processes of evaluation and judgment that had been at the heart of modern scientific organization in universities, academies, and societies since the seventeenth century. For these reasons, the Nobel awards may be seen as the crown of the tradition of positivist science.

2

Dalton's Atom and Two Paths for the Study of Matter

Dalton and the Newtonian Tradition of Mass and Force

Contemporary use of the word *atom* is a well-acknowledged misnomer at the very core of modern science. The word is a misnomer because it means "indivisible" (from Grk. *atmos*), even though the atom is now known to be a complex entity that, in the case of radioactive elements, spontaneously emits particles and radiations. Even the lightest and simplest atoms are combinations of smaller particles, with all but hydrogen containing, in addition to electrons, both neutrons and protons in the nucleus. Subnuclear particles, in turn, are composed of very small, pointlike objects called "quarks," a name borrowed from a line in James Joyce's novel *Finnegans Wake*.

The contemporary "atom," like the "molecule," belongs to both chemistry and physics, as does the man often called the founder of modern atomic theory, John Dalton. He appears in chemistry and physics textbooks as a hero of both disciplines. He adopted both Antoine Lavoisier's definition that chemical "elements" are bodies that have proven themselves undecomposable and Isaac Newton's notion that massy, indivisible particles exist that are the last, or fundamental, building blocks of matter.

Isaac Newton (1642–1727), Lucasian Professor of Mathematics at Trinity College, Cambridge, Master of the English Mint, and author of the *Philosophiae naturalis principia mathematica* (1687), defined the indivisible corpuscle very clearly in his thirty-first query of his *Opticks:* "It seems probable to me, that God in the Beginning form'd matter in solid, massy, hard, impenetrable, moveable Particles, of such Sizes and in such Proportion to Space, as most conduced to the End for which he form'd them."[1] Newton further postulated that these particles act on each other with forces varying inversely as the square of the distance between them, as he had proved to be the case for celestial bodies. He demonstrated in proposition 23 of book II of the *Principia* that a gas composed of particles repelling each other by an inverse-square force would obey Robert Boyle's law relating pressure and volume.

For many decades during Dalton's lifetime and afterwards, the words *atom* and *molecule* were used interchangeably to describe any invisible and massy particle that was capable of expressing motion and force, but the "atom" came to be particularly identified with chemical elements. Dalton's genius lay in recognizing that the hypothetical notion of the atom could give rise to a laboratory technique of measurement to determine the proportional composition in compounded substances of the chemical elements.

Like many English scientists in the eighteenth and early nineteenth centuries, John Dalton was the son of a Dissenter from the Anglican church, a Quaker who was a hand-loom weaver. Dissenters were denied entrance to the colleges at Oxford and denied graduation from the colleges at Cambridge, but the younger Dalton was fortunate in receiving some private training in science and mathematics—in particular, exposure to Newtonian philosophy. In 1793 he became professor of mathematics and natural philosophy in New College, a Dissenter-founded college in Manchester.

When the college moved to York six years later, Dalton stayed in Manchester, relishing the intellectual support and social discourse of members of the Manchester Literary and Philosophical Society, a society of entrepreneurs, teachers, clergymen, physicians, and scientific amateurs devoted to studying natural science and its applications. The society began publishing its journal, *Memoirs,* in 1785, thus providing its members with a venue for dissemination of their investigations. Dalton, who supported himself largely by private tutoring, had a close scientific friend in William Henry (1774–1836), the son of a local apothecary and author of the very successful textbook *Elements of Experimental Chemistry,* which went through eleven editions in the thirty years after its first edition of 1799.

John Dalton seated next to an air pump and sheets of figures drawn with atomic symbols. From an engraving by Worthington based on the 1814 portrait by Joseph Allen. *Courtesy of the Manchester Literary and Philosophical Society.*

By the time of his death, Dalton had achieved considerable fame not only as a provincial scientist but as one of the leading men of chemistry in Great Britain. He was praised then, as now, for the atomic theory that attached experimentally determined invariant weights to the distinct chemical elements, about fifty of which had been identified by 1814. In presenting Dalton the Royal Society's Medal for 1826, Humphry Davy, president of the Royal Society, lecturer at the Royal Institution, and the mentor of Michael Faraday, praised Dalton for fixing the proportions in which chemical bodies combine even as he held back from agreeing with Dalton on "any speculations upon the ultimate particles of matter."[2]

Modern chemists have tended to see Dalton as a successor to Lavoisier, but it was in Newton's footsteps, not Lavoisier's, that Dalton's path first began. His starting point can be found in data recorded in his notebook "Observations on the Weather," begun in 1787 and ending with an entry made the evening before his death in 1844. From this interest in the atmosphere and weather, he began studying the amounts of water vapor held in the air and concluded that, contrary to the teachings of Lavoisier and the "French school" of chemistry, water vapor could not be combined chemically with nitrogen and oxygen gases in the air, since the degree of saturation of air with water vapor varied with temperature. He also concluded that the pressure exerted by water vapor and by each gas in a mixture of gases is independent of the pressure exerted by other gases, and that the total pressure is the sum of pressures exerted by each gas.

Dalton's theory of mixed gases received a good deal of attention and some criticism, including comments from his former teacher John Gough, from the Parisian chemist Claude-Louis Berthollet (1748–1822), from Dalton's Manchester friend William Henry, and from Henry's former teacher, the Edinburgh chemist Thomas Thomson. William Henry, investigating mixtures of gases, now formulated the law that the volume of nonreacting gas dissolved in water is independent of pressure, although the weight varies directly with pressure. In a paper read before the Manchester Literary and Philosophical Society in the fall of 1803, Dalton confessed that

> I am nearly persuaded that the circumstance depends upon the weight and number of ultimate particles of the several gases: Those whose particles are lightest and single being least absorbable. . . . An enquiry into the relative weights of the ultimate particles of bodies is a subject, as far as I know, entirely new: I have lately been prosecuting the enquiry with remarkable success.[3]

Having begun with Newtonian natural philosophy, Dalton now turned, after all, to French chemistry, adopting for a starting point the table of thirty-three chemical elements from the *Traité élémentaire de chimie* (1789) by Antoine Lavoisier (1743–1794). In calculating the relative weights of the fundamental elementary substances, Dalton had to make several assumptions, one of which was the law of multiple proportions that he formulated on the basis of studies of the oxides of nitrogen. Investigating carbureted hydrogen (methane) and olefiant gas (ethylene), he concluded that twice as much hydrogen combines with a given weight of carbon in carbureted hydrogen as in olefiant gas.

Another assumption, made explicit by Dalton in lectures in 1807, was what he called his "rule of simplicity." This was the principle that when two elements form only one compound, the compound is binary—that is, that one atom of A combines with one atom of B. If two elements form two compounds, they are presumed to be one binary compound (say, AB) and one ternary compound (A_2B or AB_2). When four combinations of two elements are found, one is presumed binary, two ternary, and one quaternary (A_3B or AB_3). Measurement of vapor density of the compound decides between A_2B and AB_2, for example.

Thus, using the convention that the comparative weight of a hydrogen atom can be taken to be 1 (since hydrogen is the lightest known elementary gas), along with the law of multiple proportions and, finally, the rule of simplicity, a table of relative atomic weights can be constructed. As a starting point, water was taken to be composed of hydrogen and oxygen atoms in the ratio 1/1. Since water was found to be 85 percent oxygen and 15 percent hydrogen by weight, the atomic weight of oxygen relative to hydrogen was calculated as 5.66. The combination of nitrogen with hydrogen in ammonia (also assumed to be a binary compound) gives a relative combining weight for nitrogen. And so on.

A visit from the Glasgow chemist Thomas Thomson to his former student William Henry in August 1804 led to the first personal meeting of Dalton and Thomson, who was at the time one of the leading chemists in Great Britain. Immediately following lectures that Dalton gave in Edinburgh in 1807, Thomson published the first full account of Dalton's ideas in the third edition of his textbook, *The System of Chemistry,* largely identifying Dalton's originality with the rule of simplicity. Dalton published his own account of his work, entitled *New System of Chemical Philosophy,* the first volume appearing in two parts in 1808 and 1810, and a second volume appearing in 1827.

Dalton's "chemical philosophy" was indeed systematic, with a network of principles and hypotheses that were rooted not only in chemical experiment but also in Newtonian mechanical philosophy. He assumed that the atoms were spheres. While he later vacillated about whether atoms differed in size, in his *New System* he defended the proposition that atoms of hydrogen, oxygen, and lead differ in

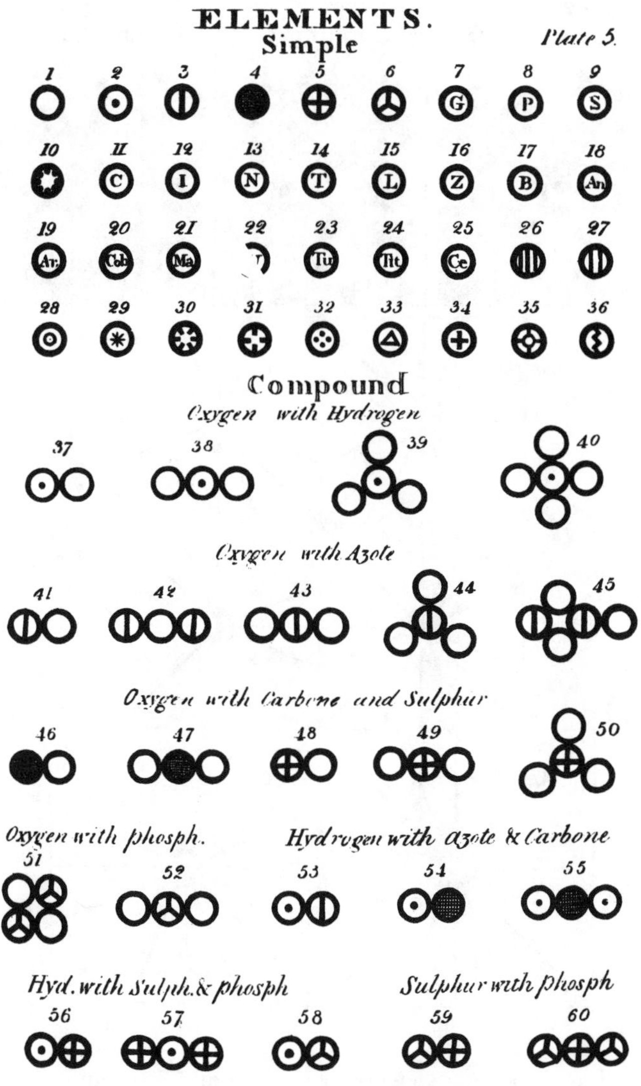

John Dalton's table of symbols for atoms and molecules in the 1810 volume of *A New System of Chemical Philosophy.* Dalton drew up lists of arbitrary symbols for thirty-six elements and suggested how atoms might be arrayed in molecules. From John Dalton, *A New System of Chemical Philosophy,* part 2 (Manchester: Russell and Allen, 1810), plate 5. *Courtesy of History of Science Collections, University of Oklahoma Libraries.*

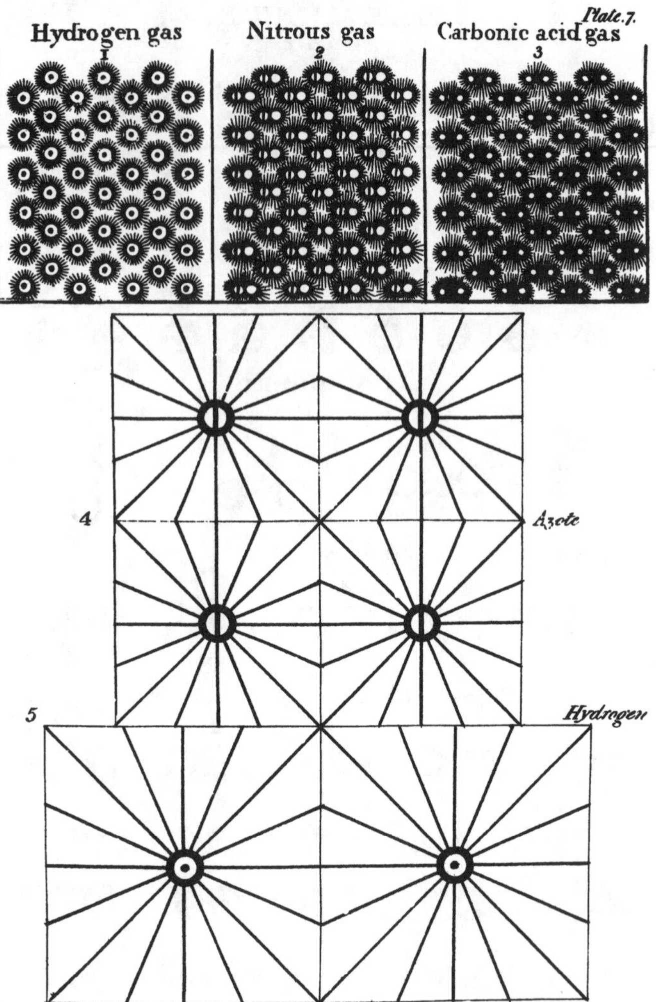

John Dalton's model for elementary gases in the 1810 volume of *A New System of Chemical Philosophy*. He assumed that gases are made up of atomic particles, each surrounded by a cloud of caloric. Repulsive forces represented by lines emanating from the atoms account for repulsion of like atoms but allow the mingling of unlike atoms where the lines do not meet. From John Dalton, *A New System of Chemical Philosophy*, part 2 (Manchester: Russell and Allen, 1810), plate 7. *Courtesy of History of Science Collections, University of Oklahoma Libraries.*

bulk just as they differ in weight. The difference in bulk meant to Dalton that at a given temperature and pressure a volume of hydrogen necessarily contains more particles, or atoms, than an equal volume of oxygen.

In Dalton's system, each atom is surrounded by a cloud of heat, which was thought of as the material fluid called "caloric." Unobserved differences in the sizes of the elementary atoms produce the observed fact that different kinds of gases form mixtures rather than layers. The reason? The heat envelopes and heat forces around the unequally sized atoms are unequal, resulting in "intestine" (not kinetic) motion. In addition, given his principle that identical atoms repel one another but dissimilar atoms exhibit neither attraction nor repulsion, Dalton argued that the rule of simplicity results from the physical circumstance that the fewer the number of like atoms in a compound substance, the greater the mechanical stability. Thus AB is the most stable, and the most common, combination of elements. On geometrical grounds, AB_{12} would represent the highest possible ratio of B to A.

For Dalton, as already noted, the most common compound of all, water, has a 1/1 ratio of hydrogen to oxygen (not 2/1, as in the modern formula H_2O; in this view, Dalton differed from some influential chemists and physicists but not from others). He represented the compound of water by two circles, one empty (oxygen) and the other with a dot in the middle (hydrogen). Dalton's symbols for compound substances combined his symbols for elementary atoms in two-dimensional space. He later had constructed for his use a collection of spheres that he could arrange in space in keeping with his notions of rules of geometrical packing.

Dalton conceived of both atoms and their combinations in a thoroughly commonsensical and naively realistic fashion. To the introduction by the Swedish chemist Jöns Jacob Berzelius (1779–1848) of alphabetical symbols for atoms (H and O, for example), Dalton acted with hostility, not only privately but in committee meetings of the section for chemistry of the British Association for the Advancement of Science. Along with William Whewell, Dalton ridiculed the algebraic appearance of symbols such as N^2O when they were first published by Berzelius.

Dalton also initially reacted with disdain and disbelief to the much-discussed hypothesis of the Italian natural philosopher Amedeo Avogadro (1776–1856) that equal volumes of gases at the same temperature and pressure contain equal numbers of particles.

Avogadro's hypothesis, like the similar one of the Frenchman André-Marie Ampère (1775–1836), relied on results of experiments conducted by Gay-Lussac, who was a protégé of Lavoisier's former associate Berthollet. Dalton regarded Gay-Lussac's law—that gas volumes combine in the proportion of small whole numbers—with skepticism, charging that Gay-Lussac was freely rounding off his numbers. There is some evidence that Dalton regarded Avogadro's hypothesis more favorably shortly before his death in 1844, although the force of his reputation lay against it.

Disciplinary Conventions: The Emergence of the Distinction between the Chemical and Physical Atom

Among the most influential English chemists in the early decades of the nineteenth century were William Hyde Wollaston (1766–1828), a Cambridge graduate who practiced medicine before setting up a chemical laboratory in London; Humphry Davy, who worked briefly for Thomas Beddoes (1760–1808) at the Pneumatic Institution in Bristol before becoming lecturer and experimenter in 1801 at the newly founded Royal Institution in London in 1801; and William Prout (1785–1850), a physician with a vigorous interest in medical chemistry and physiology. None of these men taught in the old universities of Oxford and Cambridge, where chemistry did not have an established place in the gentleman's curriculum until a good deal later in the century.

All three of these chemists objected to Dalton's hypothesis of indivisible particles. Wollaston and Davy claimed the most valuable part of Dalton's work to be the calculation of equivalent weights consistent with the quantitative and stoichiometrical approach of the German engineer and chemist Jeremias Benjamin Richter (1762–1807). Richter's work on the chemical equivalence in neutralization reactions of acids or bases to a given amount of sulfuric acid became well known in the European chemical community when it was summarized in 1802 by Ernst Gottfried Fischer (1754–1831) in his German translation of Berthollet's *Recherches sur les lois de l'affinité*.

For Wollaston and Davy, the language of "equivalents" was preferable to that of "atoms" because it avoided commitment to a speculative theory of indivisible elementary particles. Yet Wollaston chose a single equivalent value for each element, not the multiple values consistent with stoichiometrical rules, and his "Synoptic Scale of Chemical Equivalents," published in 1814, was different from Dalton's table of 1810 in only four instances.

Wollaston's table, which used oxygen rather than hydrogen as a base, was popular among many British chemists until the 1860s and inspired an 1840s version by the German professor of medicine and chemistry at Heidelberg, Leopold Gmelin (1788–1853). Wollaston's claims that his system was one of convenience and convention, rather than of theory and reality, had considerable appeal throughout the nineteenth century, although as historian Alan Rocke has noted, Wollaston could be found conjecturing occasionally about real spherical atoms.

Prout, in contrast, regarded Dalton's theory of atoms as erroneous because Prout thought he had evidence that all the chemical elements save hydrogen are compound, not simple. In a paper published in Thomas Thomson's journal *Annals of Philosophy* in 1815, Prout noted that gas densities appeared to be exact multiples of the density of hydrogen, suggesting that hydrogen is the "protyle," or basic building block, of matter. Davy, who thought he had detected hydrogen as a component of highly combustible substances such as the alkali metals, sulfur, and phosphorus, found Prout's hypothesis appealing. So enamored was Thomson with Prout's idea that he endeavored to make his experimental figures for "atomic" weights into multiples of hydrogen.

Until the mid-1830s British chemists tended to use Thomson's atomic-weight tables, which were based on theories of both Dalton and Prout, while Continental chemists used the tables of Berzelius. In 1833, Edward Turner (1796–1837), who held the first chair of chemistry at University College, London, argued vigorously in a report to the BAAS that Prout's hypothesis was unverified and that Thomson's tables were erroneous.

As already noted, the calculation of an atomic weight, or equivalent weight, depended not just on careful study of weight relations during chemical analysis but on the choice of a conventional standard for comparison (e.g., hydrogen = 1; oxygen = 10; oxygen = 100), as well as on the choice of some conventional chemical formulas (e.g., water = HO; water = H_2O). The principle of fixed proportions was successfully argued by Joseph Louis Proust (1754–1826) against Berthollet in debates in the early 1800s. Proust's success established the distinction between chemical compounds having fixed proportions and physical mixtures (including glasses and alloys) having variable proportions of components. Proust's triumph also appeared to refute Berthollet's

Newtonian argument that the mass of reactants, not only their chemical forces of affinity, might affect the direction of chemical reaction.

Other principles or laws that affected the calculation of atomic or equivalent weights were Gay-Lussac's law of combining volumes (1808), Avogadro's hypothesis (1811), and Pierre-Louis Dulong and Alexis-Thérèse Petit's law for specific heats. All these ideas emerged from the well-organized and influential school of French chemistry associated with the tradition of Lavoisier and his immediate associates and successors.

Joseph-Louis Gay-Lussac, a graduate of and then professor at the École Polytechnique, was one of a dozen or so regular guests invited on weekends to the homes of Berthollet and Pierre-Simon de Laplace (1749–1827) in the suburb of Arcueil just south of Paris. At these weekend meetings, either Berthollet or Laplace would preside over a discussion of papers that were to be presented at the Academy of Sciences on the following Monday. The weekend group at Arcueil was one means by which Berthollet and Laplace enforced a high degree of orthodoxy in Parisian science, especially with respect to Newtonian principles and inquiries in both chemistry and physics.

The Arcueil group was a mix of mathematical and experimental physicists, chemists, and earth scientists, among them Jean-Antoine Chaptal (1756–1832), Jean-Baptiste Biot (1774–1862), Étienne-Louis Malus (1775–1812), Louis-Jacques Thénard (1777–1857), Siméon-Denis Poisson (1781–1840), Pierre-Louis Dulong (1785–1838), and the German naturalist Alexander von Humboldt. Both Gay-Lussac and François Arago eventually proved to be dissenters from the ranks.

Collaboration between Humboldt and Gay-Lussac on the combustion of hydrogen and oxygen led Gay-Lussac to investigations on combining volumes of gases. In 1808, in the second volume of the *Mémoires de la Société d'Arcueil*, he published his conclusion that volume ratios in gaseous reactions are whole numbers. This result, he added, supported Dalton's new atomic theory. It also implied that, under similar conditions of temperature and pressure, equal volumes of different gases contain the same number of particles.

Dalton would have none of this, on two grounds: first, because he believed that particles of a compound gas have greater bulk than particles of an elemental gas, and, second, because experimental data showed that the density of carbonic oxide is less than that of oxygen, just as the density of steam is less than oxygen. So Dalton concluded that there must be fewer particles of carbonic oxide than oxygen and fewer particles of water than oxygen in equivalent volumes.

Within a few years, Avogadro, professor of mathematics and physics at the Royal College of Vercelli, saw a way to answer Dalton's objection. He made a clear distinction between the whole molecule (*molécule intégrante*) and the constituent molecule (*molécule constituante*), proposing that a whole molecule may consist of one or more constituent molecules. In particular, Avogadro proposed that oxygen-gas molecules consist of half-molecules (constituent molecules) of oxygen; hydrogen gas molecules consist of half-molecules of hydrogen; and the constitution of a steam molecule is two half-molecules of hydrogen and one half-molecule of oxygen. Under these circumstances, the atomic weight, or equivalent combining weight, of oxygen must be taken as twice the current value (15, rather than 7.5 at the time). Confirming experimental data lay in the fact that a volume of oxygen is about 15 times heavier than an equal volume of hydrogen.

There were some favorable reactions to Avogadro's proposal, that of Ampère notable among them, but on the whole it did not do well in the next few decades. For one thing, Avogadro was a natural philosopher, not a chemist, so that many chemists paid his physical theory little attention on the grounds that it had nothing to do with chemical equivalents. For another, Avogadro's hypothesis was often misinterpreted to mean that *every* gas molecule consists of two particles, and there was evidence that this was not the case.

A final objection was offered by a gamut of researchers ranging from Berzelius to Whewell: that identical particles could not exist in proximity as a stable entity. There were two possible, but not mutually exclusive, reasons for the instability: identical particles were thought to repel each other, either on old grounds rooted in Newton's theory of repulsive inverse-square forces or on newer grounds having to do with the electrical nature of chemical elements revealed in electrolysis experiments using the new voltaic battery.

The technique of studying gas volumes, especially vapor densities, became the tried and true war horse for analytic chemists investigating atomic or equivalent weights in the first two-thirds of the nineteenth century. The technique was perfected by Jean-Baptiste Dumas, a native of Arles, in Southern France, who began his scientific studies in Geneva and continued them under Thénard at the École Polytechnique in Paris. Regularly invited to weekly receptions at the home of the mineralogist Alexandre Brongniart (1770–1847), Dumas married Brongniart's daughter and struck up acquaintances with Arago and Thénard as well as with Brongniart's visitors

Michael Faraday and the Scottish physicist David Brewster (1781–1860).

When he was only twenty-six, Dumas adopted the hypothesis of Avogadro and Ampère with the idea of extending studies of elementary gases to vapors of elements that are not ordinarily found in the gaseous state. The simple technique he used has now been taught to undergraduates for more than 150 years.

The solid or liquid substance is put in a glass bulb. The neck of the bulb is sealed off with a torch, and the bulb is placed in a bath of water, sulfuric acid, or metal alloy and heated to a temperature sufficient to cause complete vaporization. The temperature of the vapor in the bulb is the temperature of the bath. The pressure is read on an atmospheric barometer. The bulb is then cooled and weighed to obtain the weight of the vapor, which now condenses, and the volume of the bulb is determined from the weight of water it holds. So useful and ubiquitous was the vapor-density method of determining volume weights that a French chemist in the late nineteenth century told his students that "bodies which are not volatile do not have molecular weight."

Dumas's results led him to revise some of his elders' figures for atomic weights. Assuming that the vaporized iodine molecule consists of two particles, for example, he correctly reported its atomic weight to be 127. Using the same assumption for mercury vapor, however, he incorrectly changed Berzelius's figure to half its former value. Running into difficulties with vapors of sulfur [S_6] and phosphorus [P_4], Dumas began losing confidence in the idea of the atom, telling his students in a lecture course at the University of Paris in 1836 that the word *atom* should be eliminated from scientific vocabulary.

Dumas made an important distinction, however, between the "chemical" and the "physical" atom, or what he called the *molécule chimique* and the *molécule physique*. By the late 1830s, Dumas had come to the opinion that the tables of atomic weights should be thought of, or renamed, tables of "equivalents" in recognition of the role of convention in scientific explanation. He noted the inconsistency of experimental results meant to determine the atomic or molecular makeup of elemental gases, and, indeed, he was beginning to lean more favorably toward Prout's protyle hypothesis of subunits for the chemical atom.

Not all elements could be vaporized, limiting the usefulness of the vapor-density method. After 1819, however, an important relationship between specific heat and atomic weight offered another approach. Under the influence of the Arcueil circle, Dulong concerned himself both with chemistry and physics, working as an assistant to Berthollet and then becoming professor of physics at the École Polytechnique in 1820. With his younger protégé, Alexis-Thérèse Petit (1791–1820), Dulong found that for a solid element, the product of its known atomic weight (using the convention O = 10) and its specific heat was roughly equal to 3.7. (The specific heat is the quantity of heat required to raise the temperature of one gram of a substance by one degree centigrade. For water, specific heat is 1 calorie, by convention.) Dulong and Petit called this constant number the "atomic heat" on the assumption that individual atoms have the same specific heat.

Using Berzelius's table of atomic weights as the standard, they then made some revisions, for example changing the atomic weight of lead from 2,595 (on the convention that O = 100) to a value of 12.95 (on the convention that O = 10). Berzelius, in turn, altered some of his atomic-weight values, which led to changes in formulas for a whole class of metallic oxides. Changing the atomic weight of copper, for example, from 127 (convention O = 16) to 63.5 also resulted in changing the formula for copper oxides from CuO^2 to CuO.

Mechanical Philosophy and Berzelian Chemistry

Among the defenders and proselytizers for the general themes of Dalton's atomic theory, Jöns Jacob Berzelius was perhaps the most important. Educated at the medical faculty in Uppsala, Sweden, Berzelius became an assistant in botany and pharmacy and then professor, in 1807, at the College of Medicine of the Karolinska Institute in Stockholm. In 1813 he submitted to Thomson's *Annals of Philosophy* an early version of an essay on "chemical proportions," in which he offered support for the atomic theory and proposed a new system of symbolism for the atomic elements. For Berzelius, the cause of definite and multiple proportions in chemical reactions rested in bodies that are composed of mechanically indivisible particles "that one can call *particles, atoms, molecules, chemical equivalents,* etc. . . . Once all probabilities are well considered we have every reason to represent the elementary bodies in spherical form."[4] [Emphasis original.] Berzelius did not think that Dalton's spherical symbols were very efficient, however. He suggested, instead, using the initial letter of the Latin name for each non-metal; where two or more elements had Latin names starting with the same letter, then he would use the first two letters for the metal; where the first two letters were the same, the initial letter and the first consonant not shared with another Latin name would be used. Examples in his paper were S for sulfur, Si for silicium, Sb for antimony (Lat., *stibium*), and Sn for tin (Lat., *stannum*).

The chemical sign S, for example, represents one volume of the substance, he proposed, and combinations of plus signs (+), superscripts, and parentheses might be used in the following way:

$$\text{Sulphate of copper} = CuO + SO^3$$
$$\text{Alum} = 3(AlO^2 + 2SO^3) + (Po^2 + 2SO^3)\ [Po = \text{potash}]$$

Berzelius complicated the system, in the interests of simplicity, by introducing barred symbols for double volumes (Ħ rather than H^2), superscript dots for oxygen, superscript commas for sulfur, and other symbols. Superscripts continued to be used by many chemists, especially the French, throughout the nineteenth century, although Liebig introduced the subscript convention in 1834.

Another complication was Berzelius's adoption of electrochemical ideas and electrical symbols. Like Davy, Berzelius used the new alkaline battery to decompose compound substances by electrical current, depositing their components at the electrical poles, and he came to identify the force of chemical affinity with the force of electrical attraction. His chemical formulas used grouped letters to indicate how polarized atoms or groups might be held together in a molecule or compound by slight opposite electrical charges or forces. Thus, whereas the "empirical" formula for *sal ammoniac* (ammonium chloride) would be NH_4Cl or, for *saltpeter* (potassium nitrate), would be KNO_3, the "rational" or explanatory formulas for these compounds would be

$$NH_3 \cdot HCl \quad \text{or} \quad KO \cdot NO_2$$

This electrochemical explanation of affinity came to be known as the "dualistic" theory of chemical composition and decomposition. Faraday's discovery that a given quantity of electricity liberates elements from their compounds in amounts that are proportional to their calculated equivalent weights seemed further evidence of the close analogy, if not identity, between electrical and chemical force, or affinity. Further, the dualistic theory provided ammunition against Avogadro's hypothesis: how could two atoms of oxygen form a stable molecule if oxygen ions collect at the anode (positive pole) in an electrolysis apparatus, indicating that they are negatively charged?

By the 1830s Berzelius's views were at the height of their influence, not only because of his position at the Stockholm institute, but also through his publications. His *Jahres-Bericht* (Annual Report), which reviewed recent chemical research was translated by Friedrich Wöhler from Swedish into German beginning in 1822. Berzelius's textbook *Lärbok i kemien* was translated into French and German. Among the students who came to Stockholm to study with him were some of the leading German chemists of the next generation: Wöhler, H. G. Magnus, Eilhardt Mitscherlich (1794–1863), Heinrich Rose (1795–1864), Gustav Rose (1798–1873), and Christian Gottlob Gmelin (1792–1860).

Mitscherlich's investigations of crystalline forms of arsenate and phosphate salts led Berzelius to invite him to Stockholm. Mitscherlich's law of isomorphism (1819), which stated that compounds that crystallize in the same form are similar in chemical composition, became a key aid for deciding on chemical formulas. Mitscherlich was called to the chair of chemistry at Berlin shortly after his return to Germany from Stockholm.

In 1839 Berzelius received a letter from Liebig, the former pupil of Gay-Lussac who for fifteen years had headed the most famous laboratory for training chemists in Germany, the laboratory at the University of Giessen. Liebig had used both equivalent weights and atomic weights at different times in his career, but he now wrote to Berzelius to say that he had decided unequivocally for equivalents, which in Germany had a foundation in the textbook by C. G. Gmelin's older brother Leopold, as modeled on Wollaston's work.

Why Liebig's change of heart? It seems that for some days before and after the meeting in September 1838 of the Gesellschaft Deutscher Naturforscher und Aertze in Freiberg, Liebig had driven about the countryside in a carriage with Leopold Gmelin, Wöhler, Magnus, and Heinrich Rose. He reported to Berzelius that they had found themselves in agreement to use equivalents instead of atoms: "Instead of H^2 or $\bar{H}$, simply use H. . . . Theoretical views will change, but our notation ought not. What does theory, philosophy of chemistry, have to do with notation. We will never agree on the weight of an atom, a law of nature knows no exception."[5]

Berzelius rightly replied to Liebig that the concept of the equivalent was no less conventional than the concept of the atom used by chemists: both were empirically based concepts relying on networks of preliminary assumptions and the properties of a specific series of compounds. Nevertheless, from roughly 1840 to 1900, often in the name of creating a less hypothetical, more "positive" science, the language of "atoms" often came under fire and the language of "equivalents" was often in vogue, both in Great Britain and on the Continent, especially in France.

If physical and mechanical atomism was in eclipse among many chemists in the 1840s, this was in large part due to the development and strength of a new theoretical framework within chemistry, the "structure theory." Organic chemists such as Liebig found hypotheses about combinations of groups that they called "radicals" and "types" much more useful than "atoms" in organic chemistry (see chapter five). Further, chemists found hypotheses about attractive and repulsive forces, whether "chemical" or "electrical" in nature, entirely unhelpful in predicting and classifying organic reactions. For a period from roughly 1830 to 1890, Newtonian mechanical philosophy was of little influence in chemistry; strictly speaking, it never again had a strong effect on chemical theory but was supplanted by a new physics (see chapter 6).

Mechanical Philosophy and Laplacian Physics

Similarly, in physics, Newtonian mechanical philosophy had undergone substantial revision and transformation by 1840. During the early decades of the nineteenth century, the interests of Laplace and his associates in the Arcueil group exemplified Newtonian natural philosophy in its best experimental and mathematical practice. The Laplacian program assumed the unity of natural law and natural phenomena, in the tradition of Newton, who in the preface to the first edition of the *Principia* had written,

> I wish we could derive the rest of the phenomena of Nature by the same kind of reasoning from mechanical principles, for I am induced by many reasons to suspect that they may all depend upon certain forces by which the particles of bodies, by some causes hitherto unknown, are either mutually impelled towards one another, and cohere in regular figures, or are repelled and recede from one another.[6]

In this tradition, particles of matter are characterized by mass and by their tendencies toward motion under the influence of forces that alter in strength as an algebraic function of distance.

Laplace was committed to this Newtonian vision of the nature of substance and activity. He also was committed to the practice of precise experimentation and quantification with the best instruments possible. Laplace's early (1782–1784) collaboration with Lavoisier in a series of experiments investigating animal respiration provides a striking example of this commitment. The two men confined a guinea pig for ten hours in an apparatus called an ice calorimeter. They used the amount of ice melted by the animal's body heat as a measure of the quantity of body heat. They then determined the quantity of heat and the volume of gas (carbon dioxide) produced by the combustion of a known quantity of charcoal (carbon) and calculated that the quantity of heat generated by the guinea pig, in association with a measured volume of exhaled carbon dioxide, was the amount of heat to be expected from combustion of charcoal resulting in that volume of gas. Thus they claimed to have demonstrated that respiration is a form of chemical combustion, not a mechanical process for cooling the lungs. The explanation of animal respiration as chemical combustion is a striking early instance of the program of unification through the hierarchy of the sciences as later espoused by Comte in his positivist program.

Unlike Lavoisier, who perished at the guillotine during the Revolutionary Terror in 1794, Laplace was a survivor of both the old and the new regimes. During the 1790s, Laplace was professor of mathematics at the first, short-lived École Normale Supérieure, an examiner at the École Polytechnique, and a powerful member of the Academy of Sciences. A member of the Napoleonic senate, Laplace also was briefly minister of the interior following Napoleon's coup d'état. The fruitfulness of collaboration among chemists and physicists that took place at Laplace's Arcueil estate continued to play a central role in Laplace's vision of a unified science, as he and Berthollet combined their interests and talents in laboratory investigations and salon discussions.

Laplace's *Traité de mécanique céleste* (Treatise on Celestial Mechanics) appeared during the years 1799 to 1825, verifying Newton's celestial physics and extending Newtonian mechanics to the phenomena of optical refraction and capillary action. The *Exposition du système du monde,* (1796) went even further in matters of terrestrial physics, concerning itself with the cohesion of solids, crystalline properties, and chemical reactions as a result of "molecular" forces. The authority of experiment dominated these texts no less than the authority of the differential calculus, as practiced by one of its great masters.

When it was first published, the *Celestial Mechanics* dumbfounded many British scientists, who suddenly realized the baleful irony that Newton's successors in England, who had employed Newton's style of notation instead of Continental methods, could not follow the sophisticated treatise of Laplace.

Laplace's *Celestial Mechanics* also startled the Emperor Napoleon, to whom one of its volumes was dedicated. Laplace demonstrated that Newton's law of gravitational forces was even more powerful than Newton had realized. While Newton had predicted that the system of planets and their satellites would become unstable over time, requiring the extraneous introduction of occasional forces or perturbations, Laplace showed that the average distances of the planets from the sun oscillate periodically and predictably within finite limits. According to Laplace's calculations, the solar system is completely stable and self-correcting.

Genuinely interested in science and mathematics as a young man, Napoleon was aware of the traditional view that the Creator must occasionally nudge the planets back into their orbits in order to prevent the instability that Newton had predicted. Napoleon asked Laplace why there was no mention of God in his system, to which Laplace is said to have replied with the sensational remark, "Sir, I have no need of that hypothesis."

Laplace's influence was to wane even before his death in 1827, ebbing away even as the fortunes of his patron Napoleon declined. Studies of light, heat, and electricity took non-Newtonian turns shortly after 1815 even in the bastion of Laplace's old authority, the Academy of Sciences. This does not mean that physicists relinquished Newtonian science completely, but that they amended it. Both Newtonian mechanics and the goal of unification remained central themes of the substance and rhetoric of physics.

Striking examples of the persistence of these themes come from the work of Hermann Helmholtz and Gustav Kirchhoff (1824–1887), both leaders in experimental and theoretical physics in the second half of the nineteenth century. In his 1847 paper on the "conservation of force"—said by many to provide the fundamental statement of the law of conservation of energy—the young Helmholtz wrote that all actions in nature can be reduced to attractive and repulsive forces depending solely on the distance between material points. The "solvability of this problem is also the condition of the complete comprehensibility of nature," he wrote, and the fundamental problems of natural science will be resolved once the "reduction of natural phenomena to simple forces is completed and at the same time is proven to be the only possible reduction the phenomena allow."[7]

Kirchhoff held the chair of physics at Heidelberg from 1854 to 1874 and the chair of theoretical physics at Berlin from 1875–1886. In his rector's address at Heidelberg University in 1865, he identified the goal of the natural sciences with the working out of mechanical laws that would allow us, in principle, to determine the state of matter for all future time if we knew the state of matter at any one time and all the forces acting on it. This was *precisely* the Laplacian view of the nature of Newtonian mechanics, and major physics textbooks in the mid-nineteenth century adopted the same perspective.

Corpuscles, Waves, Ether

The expressions "atom," "molecule," "corpuscle," and "material point" were all used by eighteenth and early nineteenth-century physicists for the particle that provided the mass and the coordinate points for calculations in mathematical physics. The heat that flows from hot to cold bodies, the light rays that bend as they move from air into water, and the electricity that collects at the poles of a battery were all assumed to be composed of small invisible corpuscles moving in infinitesimally small increments of time with visible results that could be predicted through the equations of the differential calculus. Heat, light, and electricity were all thought of as fluids composed of particles that were practically weightless ("imponderable") but nonetheless material. The physics practiced in this tradition was "corpuscular" or "molecular" physics.

One "imponderable" fluid, employed by René Descartes (1596–1650) in his mechanical theory and proposed by Newton in the "queries" at the end of his *Opticks,* was to play an increasingly prominent role in nineteenth-century physics. This was the peculiar fluid called the *ether.* Like Newton's later system, Descartes's had used the fundamental principles of matter and motion, but, unlike Newton's, Descartes's mechanics attributed the cause of motion to contact actions and not to immaterial force. To Newton's way of thinking, the ether was probably the material, if "imponderable," medium by which two bodies at a distance can act on each other, with the action carried by contact of intermediary ether corpuscles. Because Newton saw no means by which to demonstrate this mechanism or the ether's existence, he did not incorporate the ether into his formal, deductive mechanics. But the chimera of the ether continued to haunt physical conceptions in the eighteenth and nineteenth centuries, and it made its way back into the mainstream of natural philosophy as a result of the development of new mathematical tools as they were applied to problems of heat, light, and, later, electricity and magnetism.

The *Théorie analytique de la chaleur* (Analytical Theory of Heat, 1822) of Jean-Baptiste-Joseph Fourier (1768–1830) provided a new mathematical approach that initially seemed to undermine Newtonian physics as practiced in the Laplacian program. Fourier's approach in the early 1800s was based on the hypothesis that heat is an indestructible fluid, but Fourier did not offer a physical hypothesis for the character of the fluid. He introduced methods for mathematically treating continuously mediated actions in a fluid by means of partial differential equations, concentrating on expressions for the diffusion of heat without specifying any corpuscular or molecular model.

Both Joseph-Louis Lagrange (1736–1813), whom Fourier had assisted at the École Polytechnique in the 1790s, and Siméon-Denis Poisson, who taught mathematics and mechanics at the École Polytechnique and the Sorbonne, opposed some of Fourier's mathematical arguments. But Fourier's mathematical abilities were unquestioned. He was elected to the academy and served as permanent secretary of its mathematical sciences section from 1822 to 1830.

In his studies of heat and electricity, William Thomson later demonstrated that Fourier's equations had some important analogies with the potential functions for gravitation introduced by Lagrange in his *Mécanique analytique* (1788). Reconciliation of contact action (Fourier) and action-at-a-distance (Lagrange) reinforced mid-nineteenth-century physicists' faith in the unity of natural phenomena and natural laws, as well as in mechanical science.

In the early nineteenth century Laplace and his closest associates, especially Malus, Poisson, and Biot, applied Newtonian principles to every problem they faced in natural philosophy. Under the auspices of the Parisian academy, prize competitions were meant to elicit Newtonian treatments of the distribution of heat in solids (Fourier's subject), double refraction and polarization of light, the behavior of elastic surfaces, and the specific heat of gases. In 1817 the subject proposed for the 1819 *Grand Prix* competition was the problem of the diffraction of light. The members of the five-person prize committee were appointed, consisting of Laplace, Biot, Poisson, Gay-Lussac, and Arago.

Only two essays were submitted, one by Claude Pouillet (1790–1868), who favored the corpuscular theory of light, and the other by

Augustin-Jean Fresnel (1788–1827), a graduate of the École Polytechnique and the École des Ponts et Chaussées. Fresnel became a friend of Arago, who had entered the École Polytechnique in 1803, just one year ahead of him, and the two discussed the diffraction of light as early as 1815. Arago was one of the referees for Fresnel's first paper on diffraction published in the *Mémoires* of the academy. In 1816 Arago was in England with Gay-Lussac, and the two Frenchmen visited Thomas Young (1773–1829), learning of Young's considerable earlier work on physical optics, published in his *Treatise on Natural Philosophy* in 1807.

Reasoning from analogies with sound waves, Young had argued in a thesis at the University of Göttingen and in papers and lectures during 1802 and 1803 that a beam of light is not a bundle of asymmetric particles, as Newton had supposed, but that it is a longitudinal vibration in the medium of the ether. By this definition Young meant that particles of ether oscillate back and forth in the lines of transmission of the wave disturbance.

Young and Fresnel each demonstrated that light rays possess the property of interference, a characteristic of waves. When two or more waves are superimposed in the same medium (air, water, or ether) they either reinforce or cancel out the displacement of the medium's particles at various places, or phases, in the medium. Young demonstrated this phenomenon for light with his double-slit experiment.

Unlike interference, diffraction had been explained on the basis of the Newtonian emission theory as the result of forces of attraction. In the paper that won the *Grand Prix* of 1819, with the help of experimental apparatus provided by Arago at the Paris Observatory, Fresnel demonstrated that a wave theory of light more precisely explains diffraction, or the fringes of light seen in shadow. Poisson, who along with Laplace and Biot favored the emission theory, noted that Fresnel's equations led to the prediction that the shadow of a small round disk placed in a beam of light should have a bright spot in its exact center because of constructive interference of the light waves spilling all around the edge of the disk. Experiment confirmed this prediction.

There was one extremely important experimental property that the longitudinal theory of light had not been able to explain since Christiaan Huygens (1629–1695) had first developed a pulse, or longitudinal, wave theory in the seventeenth century. While the interference effect and what came to be called "Poisson's paradox" were inexplicable by the emission theory of light, polarization of light still required an asymmetric property in a light beam, a property that could be attributed to what Newton had called the "sidedness" of light particles. The experimental phenomenon is that the intensity of a light beam can be cut in half by its double refraction in calspar or by its reflection at prescribed angles from certain surfaces. How is this possible?

Young and Fresnel both proposed that light waves must be transverse, not longitudinal, vibrations in the ether, so that vibrations in an entire directional plane can be cut out without destroying the "beam" of light. For Young, this meant that light consisted of vibrations like the undulations of a cord secured at both ends. For Fresnel, this meant that the ether must have the properties of an elastic solid. For Fresnel and for many of his successors in mathematical physics, the major goal of physics now lay in constructing the exact properties of the ether in order to unify the physics of matter, heat, light, electricity, and magnetism.

A theory of the ether, it was thought, would be "mechanical" or "dynamical" in nature and would be capable of mathematical development. New interest among many physicists in constructing a mathematical theory of the ether based in Fourier's and Lagrange's methods drove physics into increasingly sophisticated and abstract mathematical development by the 1840s. Mathematical education in Great Britain completely changed, with men such as John Herschel (1792–1871) and William Whewell at Cambridge leading a movement for the mastery and extension of Continental methods that soon included the work of British mathematicians William Rowan Hamilton (1805–1865), George Green (1793–1841), and James MacCullagh (1809–1847). This trend in mathematical physics, however, erected an increasingly broader barrier separating the education of physicists and chemists. Ironically, the ether physics that was beginning to transform Newtonian science also was contributing to the division, rather than the unification, of chemical and physical problems.

Molecular Physics and Chemical Molecules at Midcentury

One result of this division of problems was even greater confusion than in the past over the meaning of words, a clear indication that the languages of physics and chemistry had distinct uses in the two different disciplinary communities. In the field of "molecular physics," solids, liquids, and gases were subject to mathematical description in which mathematical points and material corpuscles were identical. Molecular physics had been the goal of the Laplacian program, steeped in the Newtonian doctrine of action-at-a-distance forces, and from a mathematical point of view it had been identical to mechanical philosophy. By the mid-nineteenth century, however, molecular physics was coming to have a narrower meaning, which by the 1870s was sometimes restricted to the "kinetic theory" of gases, while theories of the ether constituted an ever-expanding field of the mechanical conception of nature.

The hypothesis of the ether continued to resist reduction to Newtonian principles of indivisible corpuscles and action-at-a-distance forces. On the one hand, Newton's old objection still held: that the hypothesis of ether particles in turn necessitated a hypothesis of a "super-ether" constituting the medium carrying these particles. On the other hand, if extremely light ether particles of very high velocities were rebounding throughout all of space as well as in the pores of matter, then one would expect, experimentally, that the heat capacity of an evacuated container is enormously larger than that of the same container filled with gas. As the American philosopher J. B. Stallo (1823–1900) noted, this is not the case. Other experiments to demonstrate the existence of the ether also failed (see chapter 3).

The tradition of molecular physics is exemplified in the work of Rudolf Clausius (1822–1888), but his work also demonstrates the very real practical difference between what nineteenth-century physicists and chemists meant by "molecules." In 1857 Clausius, then professor of physics at the Zurich Polytechnic Institute, published a pathbreaking paper entitled "Über die Art der Bewegung, welche wir Wärme nennen" (On the Kind of Motion We Call Heat). In it, Clausius introduced three assumptions about an ideal gas: that its molecules are so small that the volume they occupy can be ignored; that the duration of a collision is small compared with the time between collisions; and that the attractive forces between gas molecules are negligible. He further suggested that molecules might be thought of as complex bodies, rather than as point-masses or simple elastic spheres, since specific heat data indicate that in many gases the translational motions of molecules do not account for all the energy of the system.

Clausius appears to have thought his notion of diatomic and polyatomic "molecules" original and to have been unaware of the long-raging controversy among chemists over the Avogadro-Ampère hypothesis and what Avogadro had called in 1808 the existence of "constituent molecules." Clausius had no notion of chemists' meanings of the words "atom" and "molecule." For mid-century scientists, the failure of Dalton's program for equating the chemical and the physical atom is striking.

An important role in the reintegration of chemists' and physicists' usage was played by the Italian chemist Stanislao Cannizzaro (1826–1910), a Sicilian revolutionary who fled to Paris in 1851 to study in Michel-Eugène Chevreul's laboratory after the first unsuccessful revolt of Sicily against the Naples Bourbon monarchy. Cannizzaro fought again in Sicily in 1859, resigning a professorship in Genoa to join the forces of Giuseppe Garibaldi, and then subsequently taught at Palermo and at Rome.

In Paris, Cannizzaro learned directly of the debate still raging between equivalentists and atomists. There were those who rejected the formula H_2O and those, notably Auguste Laurent (1807–1853), Charles-Fréderic Gerhardt (1816–1856), and Charles-Adolphe Wurtz, who defended the Avogado-Ampère hypothesis and the reliability of molecular vapor densities in calculating atomic weights. Cannizzaro became not only a proselytizer for the "atomists" but also an exponent for the significance of Clausius's 1857 paper on the mechanical theory of heat, claiming that it resolved chemists' controversy over the Avogadro-Ampère hypothesis.

The vehicle for Cannizzaro's influence was his paper, published in 1858, on simplifying the teaching of chemistry by means of the Laurent-Gerhardt modification of Berzelius's atomic weights and the Avogadro-Ampère hypothesis. Cannizzaro's paper appeared in the new Italian scientific journal *Il nuovo cimento,* founded in 1854 by Carlo Mateucci and Cannizzaro's former chemistry professor Raphaele Piria. The paper's international impact was achieved through the distribution of reprints at a chemists' meeting in Karlsruhe, Germany, in September 1860 and by its incorporation into a new textbook by Julius Lothar Meyer (1830–1895), *Die modernen Theorien der Chemie* (1864). Meyer wrote that he had "received a copy which I put in my pocket to read on the way home. . . . The scales seemed to fall from my eyes. Doubts disappeared and a feeling of quiet certainty took their place."[8]

The aim of the meeting at Karlsruhe was to bring together chemists from all over Europe to discuss the need for a standardization of notation and weights. Things were worse than ever. Berzelius's barred symbols were still used by some chemists. In this case, water was represented as HO, a formula indistinguishable from

Leopold Gmelin's HO when the bars were mistakenly omitted, as often occurred in print. The use of barred symbols by Auguste Kekulé and Wurtz to represent Gerhardt's atomic weights further confused matters. Inorganic chemists tended to use equivalent notation (MO for alkali metal oxides, $C_4H_8O_4$ for acetic acid) while structural organic chemists used the atomist notation (M_2O, $C_2H_4O_2$).

During a trip to Paris in the spring of 1859, Kekulé proposed the idea of an international meeting to Wurtz. The two enlisted the aid of Karl Weltzien (1813–1870) at the polytechnic school in Karlsruhe, Kekulé's protégé Adolf von Baeyer in Berlin, and Henry Roscoe in Manchester. About 140 chemists from a dozen countries participated, responding to a series of questions about atoms, molecules, and equivalents drafted by a steering committee. Cannizzaro presented his arguments for the reform of the teaching of chemistry, and some chemists announced their agreement that the system of atomic weights designed by Gerhardt should be the basis for standardization. Others resisted the notion that scientific issues could be resolved by majority vote, and all agreed that no one was bound by the discussion.

While Cannizzaro insisted on the identity of the physical and chemical molecule, Kekulé was among those who refused to call the issue resolved, unsure whether the smallest unit of chemical reaction—the molecule—had been demonstrated to have an autonomous existence in nature. Later, in an article of 1867, Kekulé explicitly rejected the usefulness to chemists of the hypothesis of the physical atom, "taking the word in its literal signification of indivisible particles of matter." He reiterated the distinction between the "chemical atom" and the "physical atom," identifying the former with "those particles of matter which undergo no further division in chemical metamorphoses."[9]

Until the very end of the nineteenth century, some chemists refused to adopt the system of notation based on H_2O and C = 12 / O = 16, on the grounds that it involved a commitment to the principle of the real existence of the elementary atom, the identity of the chemical and the physical molecule, and the Avogadro-Ampère hypothesis. Among holdouts was Marcellin Berthelot (1827–1907), professor of chemistry at the Collège de France, French inspector-general for higher education, and permanent secretary of the physical sciences section of the Académie des Sciences from 1889 until 1907. In a debate with Wurtz and others at the academy during 1877–1878, Berthelot reiterated his opposition to what he disparagingly called the "metaphysical" belief in an indivisible atom corresponding to every elementary unit of chemical combination. He continued to write water as HO until 1891. Around 1890 the Sorbonne chemist Charles Friedel (1832–1899) could be found writing acetylene dichloride as $C_2H_2Cl_2$ for the *Berichte* of the Berlin Chemical Society, but $C_4H_2Cl_2$ for the *Comptes rendus* of the Paris academy.

One of the last important holdouts among German organic chemists was Kolbe, professor at Marburg and later at Leipzig and editor of the *Journal für praktische Chemie.* He adopted the "Gerhardt system" (C = 12 / O = 16) in 1870. The Leipzig physical chemist Wilhelm Ostwald (1853–1932) continued to fight against the appropriateness of the concept "atom" into the 1890s. In England, debates over the meaning and status of the term "atom" occurred periodically at meetings of the London Chemical Society during the 1860s and 1870s, with the somewhat eccentric Oxford professor of chemistry Benjamin Brodie (1817–1880) proposing in 1867 a "calculus of chemical operations" to be based on Boole's algebra, rather than on the hypothesis of the "atom." James Clerk Maxwell, who joined the discussion after Brodie's paper, commented that he was shocked to hear Brodie say that hydrogen and mercury are mathematical "operations." Maxwell went on to defend the identity of the physical and chemical atom on the basis of evidence from the kinetic theory of gases.

For the practical purposes of the nineteenth-century chemist, the chemical atom was a unit of reaction characterized by a unique weight, and a good deal of chemical laboratory work aimed at working out a coherent system of weights. The representation of chemical substances by symbols and formulas, however, remained a matter of considerable controversy, with many chemists believing that choosing some kinds of notation committed them to physical ideas about chemical atoms and molecules (see chapter 5).

The periodic law of the elements, which states that the properties of the elements are in periodic dependence on their atomic weights, was published in Russian by Dmitry Mendeleyev (1834–1907) in 1869 and in German by Julius Lothar Meyer in early 1870. Mendeleyev's first version of the table was presented to the Russian Physical Chemical Society, along with its basis in what he called eight fundamental principles, the first of which was the relationship between atomic weight and periodicity. Another principle noted the way in which the arrangement of the elements, or groups of elements, corresponded with their chemical combining powers or values.

Dmitry Mendeleyev first published a table of the chemical elements, listing them in vertical columns, in 1869 when he was 35 years old and a professor of general chemistry in the University of St. Petersburg. From the frontispiece of Dmitry Mendeleyev (also spelled Dmitri Mendeléeff), *The Principles of Chemistry*, part 1 (New York: P. F. Collier and Son, 1902).

Vacant spaces in Mendeleyev's table corresponded, he predicted, to undiscovered elements. A great triumph for Mendeleyev's principle came just six years later in 1875, with the discovery by Paul-Émile Lecoq de Boisbaudran (1838–1912) of what Mendeleyev had called *eka aluminum* ("eka" is from Sanskrit for numeral one). The element that Mendeleyev had calculated would have an atomic weight of 68, a density of 5.9, and an oxide formula Ea_2O_3 was found to have an atomic weight of 69.9, density of 5.93, and oxide formula as predicted. Lecoq de Boisbaudran called the new element gallium (from Lat., *gallus*, "cock," referring to its discoverer, *Lecoq* de Boisbaudran as well as suggesting Gaul).

Meyer also published an atomic volume curve, in which the periodic dependence of atomic volume (the ratio of atomic weight to density in the liquid or solid state) was clearly shown to be a function of atomic weight. The use by both Mendeleyev and Meyer of the Gerhardt system contributed to its standardization throughout Europe, although Mendeleyev disclaimed the interpretation of his periodic table as a commitment to the real existence of physical atoms of the chemical elements.

In conclusion, we see that the Newtonian program implemented in the early nineteenth century by Dalton in England, by Berzelius in Sweden and Germany, and by Laplace and Berthollet in France claimed a unity of physics and chemistry that was not realized in the course of the century. Although resistance to the identity of the chemical and physical atom, as well as to the "reality of atoms," may at first seem misplaced from a modern perspective, this is not the case. Indeed, those who resisted the simple equation of Newton's corpuscles with Lavoisier's elements, or the equation of action-at-a-distance forces with chemical affinities, provided a good deal of the impetus for experimental and theoretical investigations that were the main framework of the physical sciences in the next century.

3

The Electromagnetic View of Nature and a World of Ether

Building Theories: Fundamental Questions, Different Styles

Fundamental questions in nineteenth-century chemistry and physics were: What is matter? What are atoms, molecules, and corpuscles? But there were other central questions as well: How does matter interact with other matter? What is the role of the ether? Is electricity a substance, or is it a motion like light and heat? How can we express the properties of matter, ether, electricity, and light mathematically?

In 1867, William Thomson, who was to become Lord Kelvin in 1892, published a widely read paper, "On Vortex Atoms," in which he concerned himself with all these problems and took a firm stand against what he regarded as the old school of Newtonian and Daltonian corpuscles and action-at-a-distance forces. Writing about atoms, and tendentiously maligning chemists, he wrote,

> The only pretext seeming to justify the monstrous assumption of infinitely strong and infinitely rigid pieces of matter, the existence of which is asserted as a probable hypothesis by some of the greatest modern chemists in their rashly-worded introductory statements, is that urged by Lucretius and adopted by Newton—that it seems necessary to account

> for the unalterable distinguishing qualities of different kinds of matter.[1]

Abandoning Newton's "solid, massy, hard, impenetrable, moveable particles," Thomson announced his enthusiasm for a recent postulate of Helmholtz: that matter is an effect of vortex motion in an ether fluid, and that the investigation of the mutual action between vortex rings "is a perfectly solvable mathematical problem" that can serve as the foundation for the treatment of material corpuscles.

In Thomson's view, these vortex atoms could be the basis of a kinetic theory of gases (see chapter 4). Equally, modes of vibration of the vortex atom might account for the line spectra characteristic of the vaporized chemical elements. The period of vibration of yellow sodium light (1/525 millionth of a millionth of a second), for example, might be an approximation of the period of vortex rotation of atoms of sodium vapor. Matter, electricity, and light were all to be explained through a mathematical description of motion in the ether.

What were some of the experiments and ideas leading up to Thomson's point of view? Is his view exemplary of major trends in nineteenth-century chemical and physical science? How widely was the vortex atom accepted, and what did the vortex-in-the-ether have to do with the optical ether? We turn in this chapter to theories of ether, electricity, and light, as worked out mainly by natural philosophers from roughly 1845 to 1895.

Curiously, what this history reveals is a striking difference in preconceptions about modes of scientific reasoning among natural philosophers who were working in different national traditions. In the 1890s the French physicist, historian, and philosopher Pierre Duhem (1861–1916) made an argument for different national styles of theory-building. Similarly, in the early twentieth century, the British physicist Ernest Rutherford (1871–1937) claimed to have been struck

> by the fact that continental people do not seem to be in the least interested to form a physical idea as a basis of Planck's [radiation] theory. They are quite content to explain everything on a certain assumption and do not worry their heads about the real cause of a thing. I must, I think, say that the English point of view is much more physical and much to be preferred.[2]

There was some truth to Duhem and Rutherford's agreement that British physicists were educated to value theories that were pragmatic, concrete, and strongly visual, while, by contrast, Continental strategies of scientific explanation often were rigorous, abstract, and esoteric. To use a now-common stereotype, the British "muddled through" in order to get results that worked efficiently, while French and German scientists demanded orderly and coherent theories founded in first principles. British scientists frequently introduced the newest ideas and tools to their students, while continental scientists, especially French physicists and chemists, often believed that students should be shielded from newly speculative theories in favor of established, classical theories. In addition, Continental scientists, who traditionally received much stronger administrative and financial support from government ministries than did their British counterparts, often belittled the "amateur" and "engineering" affiliations of many British scientists.

Engineering interests were especially strong in the investigations that created the science of electricity during the course of the nineteenth century. This science of electricity was largely built on Michael Faraday's initial laboratory discoveries in the 1820s and 1830s about relationships between electricity and magnetism. From Faraday's experimental investigations there developed rival mathematical traditions of electrodynamics. One was a set of British, or "Maxwellian," theories worked out by William Thomson, J. C. Maxwell, G. G. Stokes, J. J. Thomson (1856–1940), George Francis Fitzgerald (1851–1901), Oliver Joseph Lodge (1851–1940), and Oliver Heaviside (1850–1925). Another was the "Continental" program identified mainly with the German physicists Neuman, Wilhelm Weber, and Helmholtz.[3]

By century's end, the study of electrodynamics not only encompassed the phenomena of electricity, magnetism, and light but aimed to include all physical and chemical phenomena in an electromagnetic theory of the ether. In the 1880s Helmholtz's student Heinrich Hertz (1857–1894) sought a test of rival electromagnetic theories and ended by tipping the balance against Helmholtz and toward Maxwell by creating the electric waves, afterwards called radio waves, that had been predicted by Maxwell.

Experiments to detect the ether, however, continued to fail at century's end, even while Joseph Larmor and Henrik A. Lorentz (1853–1928) worked out theories that turned out to be the last great expressions of the electromagnetic view of nature rooted in the ether. By the 1920s (as we will see in chapters 6 and 7), the "electron" proposed by Larmor and Lorentz no longer needed the ether as its seat.

Ironically, given the ether-bound history of electromagnetic theories during the course of the nineteenth century, Faraday himself abandoned the ether hypothesis toward the end of his career—just when most other scientists were coming to think they could hardly imagine a world without it.

Faraday the Nonconformist in Chemical and Natural Philosophy

The story of Michael Faraday is a success story in the up-by-the-bootstraps tradition that has been part of the mythology of self-help and individualism in English national life. Like Dalton, Faraday has been claimed as a hero for both chemistry and physics, although he preferred the title "natural philosopher."

The son of a blacksmith, Faraday was apprenticed to a bookseller, stationer, and bookbinder when he was thirteen years old. Thanks to the beneficence of a shop customer, young Faraday received a ticket to one of Humphry Davy's chemical lectures at the Royal Institution. Faraday was precisely the sort of "working man" for whom the Royal Institution had been founded in 1800, and he turned out to be its greatest triumph.

After writing out the text of four of Davy's lectures and employing his skills to bind them, Faraday presented himself to Davy and was fortunate in soon being offered a position as Davy's assistant. The following year, in 1813, Faraday accompanied Davy and his wife on a trip to the Continent during a period when England was at war with France. Despite the war—and despite his personal lack of gentility—Faraday found himself on a gentleman's tour; although he was in a subordinate capacity, he nonetheless met men of science who were Davy's peers. Among these were the Geneva natural philosopher Gaspard de La Rive (1770–1834) and his son Auguste-Arthur (1801–1873), who became Faraday's lifelong friends and correspondents.

On his return to England, Faraday became assistant to Davy's successor, William Brande (1788–1866), and continued to collaborate with Davy on projects. In 1825 Faraday, now a Fellow of the Royal Society, became director of the Royal Institution Laboratory, and in 1833 he was appointed the first Fullerian Professor of Chemistry. His laboratory—now a museum display—remains one of the showpieces of the Royal Institution.

Faraday's own research provided the subject for many of his public discourses, which included the Royal Institution's Friday Evening

Discourses and the Christmas-season "juvenile" lecture series largely created by him. The techniques used in lecture demonstrations and in the laboratory were discussed and illustrated in Faraday's *Chemical Manipulations* (1827). By 1827 he had many chemical investigations to his credit, from one published in 1816 on "caustic lime" to his isolation and identification of properties of "bicarburet of hydrogen" (benzene) in 1825.

Like Davy, Faraday was suspicious of Daltonian atoms and preferred to avoid the language of atoms; also like Davy, he inclined toward the view that the power of chemical affinity is associated with the power of electricity. As chairman of the chemistry section of the British Association for the Advancement of Science in the 1830s, Faraday was to direct much of the section's attention to studies in electricity. One of his greatest contributions to chemistry was the demonstration that a fixed, measurable quantity of electricity flowing from a voltaic battery characteristically releases from water 1 gram of hydrogen and 8 grams of oxygen, or, in general, the chemically established equivalent weights of elements decomposed from solution and deposited at the poles of the electrolytic cell. Here was independent evidence from chemical analysis for the concept of chemical equivalents.[4]

On the face of it, this result might seem supportive of Berzelius's theory of electrochemical dualism, discussed in chapter 2. In fact, Faraday undermined Berzelius's theory by demonstrating that the quantity of electricity in electrolytic decomposition did not depend at all on degree of "affinity." While Faraday conceded that the vocabulary of the atomic theory was consistent with his electrochemical results, and indeed that his experiments supported Dalton's numbers for atomic weights, he nonetheless concluded, "I must confess I am jealous of the term *atom;* for though it is very easy to talk of atoms, it is very difficult to form a clear idea of their nature."[5]

By the early 1830s, an increasing amount of Faraday's time was taken up with experimental studies of electrochemistry and electromagnetism. The electromagnetic experiments originated in an assignment Faraday accepted for the *Annals of Philosophy.* He was to write an account of Hans Christian Oersted's discovery that a magnetic needle twists circularly when placed above or below a platinum wire through which a current passes.

Both William Wollaston, who was then president of the Royal Society, and Davy were much interested in this discovery, for which Oersted received the Royal Society's Copley Medal in 1820. Wollaston tried an experiment with Davy to make a wire rotate on its axis when the wire carried a current in a magnetic field, but they found no positive result. Faraday successfully carried out a similar experiment a few months later, in the fall of 1821, without, as it happened, giving what some of Wollaston's friends thought should have been proper credit to him. In Faraday's "electromagnetic rotator," a pivoted bar magnet rotated around a wire-carrying current and a pivoted wire-carrying current, reciprocally, rotated around a bar magnet.

All through the 1820s experiments in electromagnetism were popular in London, as elsewhere, among both scientific amateurs and professionals. In lectures given by both George Birkbeck and William Sturgeon, a model of the earth's magnetic properties was exhibited in the form of a grooved wooden ball around which wires carrying electricity were wound. The point of the model was to show that magnetism could be caused by electric currents flowing in curves.

From 1831 on, Faraday hardly left the subject of electromagnetism. That year, he set himself the following problem: if magnetism could be produced from moving electrical currents, could electricity be produced from moving magnetic power? After a number of trials, recorded in notebooks that were later published, he found that when current flows intermittently from a battery through coils of wire wound around one side of a soft iron ring, with the wire insulated from the iron by calico cloth, electric current appears in a second coil wrapped around the other side of the ring and connected to a galvanometer. He soon found that the iron ring was unnecessary to produce the effect.

There followed experiments demonstrating that electricity is produced by moving a bar magnet in and out of a coil of 220 feet of copper wire wrapped around a paper cylinder and that electricity is generated in a copper disk or a copper coil rotated between the poles of a magnet.

Gradually, as Faraday carried out thousands of experiments (a total of sixteen thousand numbered entries in his published laboratory notes), he developed a full-blown theory of electromagnetism clearly distinct from Newtonian theory of action-at-a-distance forces. The forces that Faraday was studying acted along curved lines in a three-dimensional field, not in straight lines across the shortest distances between points. Newtonian forces of electrical attraction and repulsion had to be derived from the field; they did not constitute the field.

A crucial test for Faraday's notions came in experiments undertaken during 1845 and 1846—experiments he was urged to do by the young William Thomson, whom Faraday met for the first time at the September 1845 meeting of the BAAS in Cambridge. The experiments resulted in the discovery of what came to be called the Faraday effect.

The Faraday Effect and Its Implications

Experiment had shown Faraday that electromagnetic effects are transmitted in curved lines and that a charged body can induce charge on a second body despite the obstacle of a nonconducting shield. Thus, he thought that either the ether or the medium of space itself must be filled with curves or lines of electric or magnetic force stretching like taut elastic threads from one charged surface to another or from one magnetic pole to another. These curves could be mapped with iron filings around a horseshoe or bar magnet (his first drawings of the lines were published in 1851). They are like lines of latitude and longitude, or the lines of electricity in Birkbeck and Sturgeon's sphere.

For Faraday, the wires along which current runs do not contain electricity but rather *conduct* it, so that the electricity, or electrification, is on the surface of the conductor. This he proved by entering a huge conducting cube, showing that the charge that is conferred on the cube has no influence inside the cube. The insulating substance (dielectric) between conductors is in a state of strain, so that, as he described it in the 1830s, the dielectric's particles are polarized like a series of small magnetic needles. Different dielectrics have different specific inductive capacities or powers. Any change in the tension of the "electrotonic" state of the medium in which the curves of force exist always gives rise to the production of electromotive force.

Faraday's theory seemed to presume an ether, the imponderable substance already at the heart of mathematical optics. Would a transparent dielectric—Thomson inquired of Faraday in 1845—have an effect on light? If the ether medium between electrical or magnetic poles is in a state of strain, would there not be an effect on a beam of polarized light, since optical theory predicts that an asymmetrical medium rotates polarized light?

This was not an entirely new idea to Faraday, who in 1833 had unsuccessfully sought an effect on a beam of polarized light as it passed through a solution undergoing electrolysis. He now repeated his earlier experiments, still not finding an effect. (It eventually was found in the mid-1870s by John Kerr, who used more sensitive apparatus.) After trying different arrangements for passing polarized light through various transparent materials (such as flint glass and rock crystal), Faraday finally discovered a positive effect with a piece of lead glass when the light was transmitted along the lines of magnetic force for a particular orientation of the magnetic poles.

The implications of this experiment were startling. For mathematical physicists, here was positive confirmation that the ether was subject to rotational stresses and strains in an electrical or magnetic "field," a term first used by Faraday in 1845 and first published in 1846. Here, along with the phenomena of dispersion, absorption, and stellar aberration, was another property of the ether that had to be taken into account in constructing a fluid or elastic-solid theory of the ether.

Indeed the need for an ether seemed all the more pressing as reports were published by the French experimentalists Hippolyte Fizeau (1819–1896) and Léon Foucault (1819–1868). During the period from 1849 to 1862, Fizeau and Foucault demonstrated that the speed of light is greater in air than in water, in exactly the ratio predicted by Fresnel. These experiments provided conclusive proof for the wave theory of light against the rival Newtonian particle theory.

Paradoxically, for Faraday the ether was becoming less of a reality. In an article on ray vibrations in 1846 he dismissed the distinction between ponderable matter and imponderable ether, saying that the difference between them could only be one resulting from the numbers of lines of force, not kinds of particles. In 1852 he published articles on lines of magnetic force in the *Philosophical Transactions of the Royal Society* and in the *Philosophical Magazine* in which he laid out how customary laws of action-at-a-distance may be derived from the lines.

According to Faraday, each line corresponds to a unit of magnetism or electric charge. The lines contract or expand, accounting for rectilinear attraction and repulsion phenomena, and they thin out as the square of the distance from the central axis running between opposite charges or magnetic poles. But he rejected the need for a special medium in which to embed the lines.

Now, as in his earlier "Speculation Touching Electric Conduction and the Nature of Matter" (1844), Faraday emphasized the notion that matter is known by its agency of power and action, and he deemphasized his own earlier notions of contiguous particles, material polarities, and imponderable ether. In identifying matter with power, he made reference to ideas put forward by the Serbo-Italian natural philosopher Ruggero Giuseppe Boscovich (1711–1787) in the late eighteenth century, but this reference may reflect convention rather than real debt.

Much has been written about the possible influence on Faraday of the philosophical traditions of *Naturphilosophie,* perhaps transmitted through Humphry Davy and the poet Samuel Taylor Coleridge, and of the religious views of the nonconformist Sandemanian (Church of Christ) sect to which Faraday belonged. Yet, as is well known among students of the history of nineteenth-century science, the notion of the interconvertibility of forces was a common topic that was "in the air" in the 1840s. As will be discussed in Chapter 4, this was true for Helmholtz and for J. R. von Mayer in Germany and for James Joule and William Thomson in England.

As Faraday put it in 1845, "the various forms under which the forces of matter are made manifest have one common origin; . . . they are convertible . . . one into another, and possess equivalents of power in their action."[6] The seriousness with which Faraday's younger colleagues were to take his ideas and integrate them into the mainstream of the Cambridge tradition of mathematical physics shows not only the power of his experimental demonstrations but the resonance of his non-Newtonian presumptions with similar notions in the intellectual milieu of the time.

The Cambridge Approach to Faraday's Natural Philosophy

As the historian of science Ole Knudsen has remarked, it was William Thomson who in the early 1840s discovered a mathematical equivalence of theories based on mutually exclusive physical concepts: Newton's concept of action at a distance and Faraday's concept of action propagated in a field. Thomson's presentation of this mathematical equivalence had at least two very important results. First, it made Faraday's experimentally-based principles respectable among mathematical physicists. Second, it demonstrated the usefulness of mathematics in the actual construction of physical theories as well as in expressing known physical laws. If two mathematical formulations could be demonstrated to be analogous in their structure or equivalent in their predictive outcomes, then new insight into the underlying physical phenomena represented by the two mathematical systems could be obtained.

In strong contrast to Faraday, Thomson was a member of a privileged academic family. His father, James Thomson, became professor of mathematics at Glasgow University in 1832. Educated at Glasgow

University and then at Saint Peter's College in Cambridge, William Thomson had mastered Joseph Lagrange's *Mécanique analytique* and Jean-Baptiste Fourier's *Théorie analytique de la chaleur* by the age of fifteen. When he was only eighteen and a recent arrival in Cambridge, Thomson published a paper (his third published paper) in the *Cambridge Mathematical Journal.* This paper, "On the Uniform Motion of Heat in Homogeneous Bodies, and Its Connection with the Mathematical Theory of Electricity" (1842) was a starting point for developing mathematical representations of Faraday's electrostatics and electrodynamics.

In Paris after he left Cambridge, the twenty-one-year-old Thomson made the acquaintance of French mathematicians including Joseph Liouville (1809–1882), the editor of the *Journal de mathématiques pures et appliquées.* The young Thomson's expertise inspired Liouville to ask him to write a paper on Faraday's electrostatics for a French audience, a paper that appeared in 1845.

At this time, the Continental approach to electricity lay largely in the tradition of French mathematical physics, especially the work of Poisson and Ampère, who along with Arago assimilated the phenomena of electromagnetism into an action-at-a-distance framework. Ampère, who like Faraday concerned himself with chemical as well as electrical phenomena, proposed the notion of an electrodynamic molecule in which magnetic effects result from circular electric currents within matter. Ampère's mathematical treatment of relations between two electric currents assumed that forces acting rectilinearly give the appearance of circular force.

Thomson's 1842 paper approached matters differently, however, developing an analogy between electrostatic phenomena and phenomena caused by the uniform flow of heat from one part of the solid body to other parts or to the body's surroundings. As J. C. Maxwell later put it, "The similarity is a similarity between relations, not a similarity between the things related."[7] The analogy is expressed in table 1.

Thomson's demonstration of how Faraday's curved lines of force correspond to lines of heat flux effectively correlated Faraday's forces with action-at-a-distance forces by means of mathematical representation. The differential equation that Thomson constructed could be used to represent equilibrium of temperature, attraction of bodies, or motion of a fluid.

The mathematical analogy was limited by the assumption in Fourier's theory of a steady flow of heat across an interface, while in

Table 1 William Thomson's analogy between electricity and heat.

Electrostatics	Heat
The electric field	An unequally heated body
A dielectric medium	A body that conducts heat
Electric potential at different points in the field	Temperature at different points in the body
EMF tending to move positively charged bodies from places of higher to lower potential	Flow of heat by conduction from places of higher to lower temperature
A conducting body	A perfect conductor of heat
Positively electrified surface of conductor	A surface through which heat flows into the body
Negatively electrified surface of a conductor	A surface through which heat escapes from the body
A positively electrified body	A sink of heat, i.e., a place at which heat disappears from the body
An equipotential surface	An isothermal surface
A line or tube of induction	A line or tube of flow of heat

Faraday's theory discontinuous fluxes occur from one medium to another. Thomson ignored this discrepancy, however. Thomson's approach supposes the conservation of heat and of electricity, like all the forces of nature. As was clear in both the 1842 and 1845 papers, he treated electricity as a substance (fluid) within the dielectric.

In 1846, the year that Thomson, age twenty-two, took up the chair of natural philosophy at Glasgow University, he published another important paper. In order to represent the magneto-optic rotation confirmed by the "Faraday effect," Thomson sought a mechanical representation of electric, magnetic, and galvanic forces. In this paper Thomson introduced the operator *curl*, that is, the differential rotation of a volume element of a solid medium about the axis of a magnet:

$$\vec{F} = \mathrm{curl}\,\vec{A}$$

This work helped to establish in British electromagnetic theory the notation of "vector potential" for magnetic force.

Thomson wrote to Faraday that this paper was only a "sketch" but that he thought that a detailed mathematical theory could be developed that would explain the effect of magnetism on polarized light, bringing together the phenomena of light, electricity, and magnetism. As he corresponded and collaborated with George Stokes, Thomson became increasingly interested in mathematical theories and visual models relating ether to matter, as in the statement on "vortex atoms" quoted at the beginning of this chapter.

Thomson's interest in vortex and hydrodynamics models was also stimulated by the Edinburgh engineer and physicist William Rankine (1820–1872), who proposed in 1850 that heat is the result of rotational motion of atmospheres about "motes," or molecules, of matter. Along with Stokes, Thomson began developing strongly visual, even tactile, mechanical models of the ether, or what he in the 1850s began calling the "air-ether" or "aer".

This use of mechanical models by Thomson and many of his British colleagues was based not on astronomical models, as was common on the Continent, but in experience of ordinary materials and engineering practice. Thomson, for example, spent much time as an engineering consultant for the laying of trans-Atlantic telegraph cables during the 1850s and 1860s. His demonstrations and research with students in his Glasgow laboratory were oriented as much toward resolving engineering and practical problems as toward solving theoretical problems.

In their biographical study of Lord Kelvin, Crosbie Smith and M. Norton Wise have called his method of modeling a "look-and-see" method that emphasized sensory perception—feeling and touching—to know cause and effect as a "potentially real thing." In lectures given in Baltimore in 1884, in order to help them think about the interaction of a molecule with the ether, Thomson encouraged students to embed a ball in a bowl of jelly; "Apply your hand," he said, and "produce vibrations in your jelly solid by taking hold of this ball and shoving it to and fro." To show coupled vibrations inside a molecule, Thomson constructed what he called a "wiggler," consisting of a set of weighted wooden bars attached to a piano wire and suspended from the ceiling. A box strung with cords was used to demonstrate stress and strain in the ether.[8]

These devices, like the many devices described in Oliver Lodge's *Modern Views of Electricity* (1889), comprised an imaginative physical "cabinet" of British natural philosophy— ridiculed by Pierre Duhem in a review of Lodge's book:

Lord Rayleigh (John William Strutt) and Lord Kelvin (William Thomson) in Lord Rayleigh's laboratory at his private residence at Terling in Essex in July 1900. In an aristocratic and gentlemanly tradition, Rayleigh and Kelvin helped establish standards for precise laboratory instrumentation and measurement through collaborations with researchers in their own private laboratories, as well as in laboratories associated with universities. From plate 53 in Robert Andrews Millikan, Duane Roller, and Earnest Charles Watson, *Mechanics, Molecular Physics, Heat, and Sound* (Cambridge, Mass.: MIT Press, 1937) reproduced from Robert John Strutt, *John William Strutt, Third Baron Rayleigh, O.M., F.R.S.* (London: Edward Arnold and Co., 1924; new ed., Madison: University of Wisconsin Press, 1968). *Courtesy of MIT Press and University of Wisconsin Press.*

> Here is a book intended to expound the modern theories of electricity and to expound a new theory. In it there are nothing but strings which move around pulleys, which roll around drums, which go through pearl beads, which carry weights; and tubes which pump water while others swell and contract; toothed wheels which are geared to one an-

> other and engage hooks. We thought we were entering the tranquil and neatly ordered abode of reason, but we find ourselves in a factory.[9]

Maxwell, Electromagnetism, and Methodology in Theoretical Physics

Equally alarming to Duhem was James Clerk Maxwell's 1873 treatise on electricity and magnetism, which Maxwell wrote as a chronological and experimentally based account of his investigations and theories since 1856. The book seemed to many readers a badly organized, incoherent assemblage of experiments, derivations, equations, and theories—a far cry from the Continental fashion of laying everything out according to first principles.

Maxwell himself advised the reader to approach the four parts of the *Treatise* in parallel readings rather than in sequence. Further, his rejection of the notion of an "electrical fluid," his introduction of a "displacement current," and his long list of equations for the electromagnetic field, stated both in full Cartesian coordinates and in quaternions, provided for difficult and often incomprehensible reading.

The book appeared shortly after Maxwell had been appointed professor of experimental physics at Cambridge University and first director of the Cavendish Laboratory. By this time, Maxwell had a considerable reputation as a natural philosopher and mathematician. Maxwell was from a prominent Scottish family and entered the University of Edinburgh at sixteen, before going to Saint Peter's College in Cambridge. Another Scotsman, Peter Guthrie Tait (1831–1901), who was Thomson's coauthor for the *Treatise on Natural Philosophy* (1867), was then a senior student at "Peterhouse," as the college was called. Tait was to win out over Maxwell in 1859 in competition for a chair of natural philosophy at Edinburgh.

Following in his fellow Scotsman William Thomson's footsteps, Maxwell finished as second wrangler in the mathematical tripos at Cambridge. He triumphed on the 1854 Smith's Prize Examination, one of the questions being a proof of the "Stokes Theorem," which appeared for the first time in print on this exam. This theorem states the equality between the integral of a vector function around a closed curve and the integral of its curl over the enclosed surface. It was to have important applications in electromagnetic theory.

Maxwell returned to Scotland in 1856 in the chair of natural philosophy at Marischal College in Aberdeen before becoming professor of physics and astronomy at King's College in London from 1860 to 1865. The next years were spent at his family home in Glenlair with frequent trips to Cambridge, where he served as an external examiner. It was in these years that he wrote the *Treatise on Electricity and Magnetism.* His wife, Katherine Dewar Maxwell, was the daughter of the principal at Marischal College and worked with her husband during the London and Glenlair years in experiments on color vision and kinetic theory.

The appointment to the Cavendish Laboratory returned Maxwell to Cambridge. The laboratory was founded in 1870 through a generous gift from the seventh duke of Devonshire, who was chancellor of the University and a descendant of the eighteenth-century chemist and natural philosopher Henry Cavendish. Maxwell was the third choice for the directorship after Thomson and Helmholtz each declined to move to Cambridge. Maxwell's experimental and theoretical work fell off after 1871, with much of his time devoted to editing the electrical manuscripts of Henry Cavendish and to setting up the Cavendish as a laboratory of precision electrical measurement. Indeed, there was much hostility to the laboratory at the University of Cambridge because of its industrial appearance and its routinized training program for postgraduate students.

The Stokes theorem was one of many mathematical techniques that Maxwell applied to electricity and magnetism. His university studies, as well as his early correspondence with Thomson, led him in 1856, as he took up his youthful post in Aberdeen, to publish a paper that built on Faraday's and Thomson's work. Maxwell's paper "On Faraday's Lines of Force" had the purpose to "show how by the ideas and methods of Faraday, different types of phenomena which may be discovered may be made mathematical." The method was one of analogies: "In order to obtain physical ideas without adopting a physical theory, we must make ourselves familiar with physical analogies. . . . A partial similiarity between the laws of one science and those of another . . . makes each of them illustrate the other."[10]

Maxwell concerned himself with analogies between the formulas for heat, electricity, and gravitation. He proposed an imaginary fluid, "a collection of imaginary properties which may be employed for establishing certain theorems in pure mathematics." This fluid was incompressible, in steady motion, and arranged in tubes corresponding to Faraday's lines of force. Maxwell then developed analogies between electricity and magnetism.

In his formulation, the quantity of magnetism in a body is equivalent to the number of lines of magnetic force that pass through the body. The intensity of magnetism depends on the resisting power of a volume section as well as on the number of lines through the section. By the use of vectors corresponding to electric and magnetic "intensity," or force (*E* and *H*), and electric and magnetic "quantity," or flux (*I* and *B*), equations can be derived for induced electromotive force.

This force is proportional to the change in the number of lines of inductive magnetic action passing through a circuit and to the intensity and direction of change of state in the magnetic field. Faraday, Maxwell noted, referred to this state as the "electrotonic state": a state into which all bodies are thrown in the presence of magnets and currents.

The theory of 1856, he concluded, is a possible alternative to Wilhelm Weber's theory of positive and negative electrical masses moving through a conducting wire, although Weber's is a better *physical* theory. "[It] is a good thing to have two ways of looking at a subject and to admit that there *are* two ways of looking at it," Maxwell notes, adding "I hold that the chief merit of a temporary theory is, that it shall guide experiment, without impeding the progress of the true theory when it appears."[11] This statement by Maxwell was a powerful argument for the value of the heuristic approach in the construction of scientific theories. A hypothesis or theory that is not physically true might be used legitimately on the pathway to a better theory.

In 1861 Maxwell arrived at a theory that predicted novel experimental results. He proposed two new physical phenomena that were potentially testable: the "displacement current" and electric waves moving at the speed of light. The argument of the paper is clear and brilliant.

The distribution of iron filings in the vicinity of a magnet, begins Maxwell, makes us think that lines of force are real and exist even when there is no magnet present. Suppose that the phenomena of magnetism depend on tension in the direction of lines of force, combined with a kind of hydrostatic pressure—that is, a pressure greater in the equatorial than in the axial direction. What mechanical explanation can we give of this inequality of pressure in the fluid?

We may suppose that the mechanical origin lies in molecular vortices, with the axes of the vortices parallel to the lines of force. The vortices are presumed to be small in size in comparison with molecules (i.e., ordinary matter). We know that the lines of force are affected by electric currents and that they are distributed about a current. But what is the physical connection between the vortices and the currents?

Continuing to draw on mechanical conceptions, Maxwell conceived of tiny idle wheels rotating in the medium between each pair of vortex cells, with the wheels rotating in place in the direction opposite to the rotation of the vortex cells. If the rotational velocity of a vortex cell changes with respect to an adjacent one because of strain in the medium, then translation of the idle wheels is caused, accounting for electric current.

The rotation of the idle wheels transmits the motions of the vortices from one part of the field to another, and the tangential pressures called into play constitute electromotive force. In a dielectric or insulator, the idle wheels do not translate, but are only slightly displaced because of strain in the medium. Thus, in a nonconductor, displacement causes electric polarization and a change in displacement causes an extra current (displacement current) in addition to the usual conduction current.

After making some assumptions about the elastic properties of the vortex medium, Maxwell calculated the velocity of transverse elastic waves that is caused by propagation of an electric displacement in the medium. In doing this, he concluded that the waves would move at a speed equal to a constant value: the ratio of the (electrostatic) force exerted by an electric charge on another charge to the (electrodynamic) force exerted by an electric charge on a neighboring magnetic pole when the electric charge flows through a conductor.

Gustav Kirchhoff had concluded in 1857 that the ratio of electrostatic to electrodynamic charge has the dimensions of a velocity and is of the same magnitude as the speed of light. In 1855 Wilhelm Weber and Rudolf Kohlrausch had published experimental results that this velocity is 3.11×10^{10}, which is similar to Fizeau's value of 3.15×10^{10} for the speed of light. However, they explicitly denied a similarity of origin. In contrast, Maxwell concluded,

> The velocity of transverse undulations in our hypothetical medium, calculated from the electromagnetic experiments of MM. Kohlrausch and Weber, agrees so exactly with the velocity of light calculated from the optical experiments of M. Fizeau, that we can scarcly avoid the inference that *light consists in the transverse undulations of the same medium which is the cause of electric and magnetic phenomena* .[12] [Emphasis original.]

Maxwell viewed his mechanical theory of vortices as a temporary and provisory hypothesis, and, following up on the methods that Thomson and Tait were developing in their *Treatise on Natural Philosophy,* he moved within the next few years to free his electromagnetic theory from physical hypothesis and ground it in the dynamical formalism of Hamiltonian and Lagrangian functions, that is, principles of action and energy. His "Dynamical Model of the Electromagnetic Field"(1864) roots electromagnetic phenomena in the kinetic energy of motion of parts of the ether and in the potential energy residing in the structure ("elastic resilience") of the ether without specifying the motion or structure. The propagation of undulations in this medium, Maxwell argued, is the continual transformation of potential energy into kinetic energy.

Did Maxwell's route to his final equations of electromagnetic theory really mean that he relinquished the hypothesis of a material ether? Hardly so. In the 1870s he expressed confidence that there is a material medium for light and electricity, while still rejecting the notion of a material "electric fluid."[13] In an often-quoted article on the "Atom" in the *Encyclopaedia Britannica* of 1875, Maxwell praised the notion of the vortex atom and commended its mathematical solution as one that would bring great glory to its solver. In his *Encyclopaedia Britannica* article on "Ether" (1879), he suggested an earth-based method of detecting an ether wind by measuring variations in the velocity of two beams of light making journeys back and forth between two different mirrors, but he concluded that the effect would be to small to detect.

As for experimental confirmations of Maxwell's electromagnetic theory, there were several possibilities. One was a proof of electromagnetic radiation pressure, a prediction of the theory. As we will see in the last section of this chapter, William Crookes initially thought he had this proof in 1875. Another possibility was confirmation of the displacement current and of electromagnetic waves moving at the speed of light, as discussed in the next section of this chapter. A third possibility was discovery of phenomena further undermining the hypothesis of an electric fluid and consistent with Maxwellian principles.

As historian Jed Z. Buchwald has noted, this last proof was the considerable, but short-lived, achievement of two Americans, Edwin Hall (1855–1938) and Henry Rowland (1848–1901). In 1879, at the time he discovered what came to be called the Hall effect, Hall was studying Maxwell's *Treatise* under the direction of Rowland, the di-

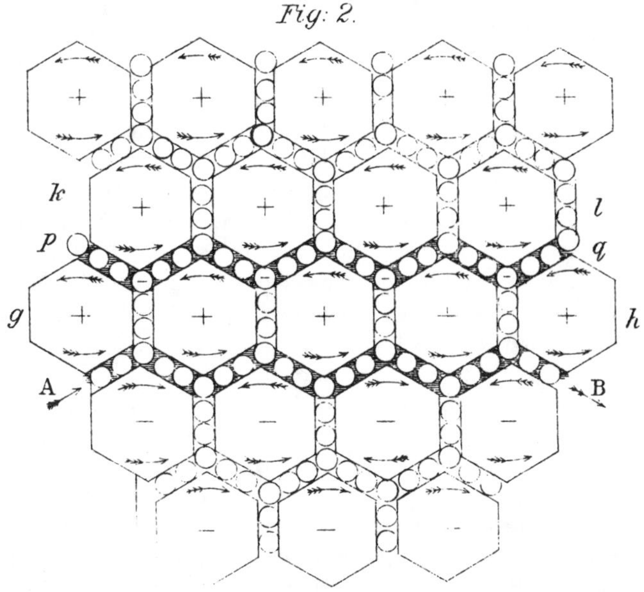

James Clerk Maxwell's mechanical model of electrical particles and molecular vortices. The motion of moveable particles (AB, *pq*) between neighboring vortices (*gh, kl*) is taken to be an electric current. As this current begins from left to right at AB, the row of vortices *gh* is set in motion in the opposite direction. Electric current then is induced in the layer of particles *pq* as the rotatory velocity of the vortices is communicated from one part of the field to another. From James Clerk Maxwell, "On Physical Lines of Force. Part II. The Theory of Molecular Vortices Applied to Electric Currents", pp. 467–88, plate 8, following p. 488 in *The Scientific Papers of James Clerk Maxwell,* edited by W. D. Niven, vol. 1 (Cambridge: Cambridge University Press, 1890). *Courtesy of Cambridge University Press.*

rector of the new physics laboratory at the Johns Hopkins University in Baltimore. In Hall's experiment, an electric current was sent across the length of a metal (for example, copper) plate that was placed between the poles of an electromagnet, with the field normal to the plane of the plate. A galvanometer attached across the plate's width detected a current only while the field was turned on.

Hall and Rowland interpreted this phenomenon to mean that an electric current and a subsidiary electric field were produced in the magnetic field, and in 1880 Rowland incorporated Hall's new "field" into one of Maxwell's equations (the "Faraday law" for magneto-optic rotation). The new version of the equation led to a wave equation that nicely explained the Faraday effect. Richard T. Glazebrook (1854–1935) at Cambridge then took up Rowland's treatment and demonstrated that the Hall term that Rowland had introduced could in fact be derived from Maxwell's equations in the *Treatise*. Many, certainly Rowland, saw this as a vindication of Maxwell's theory, although the effect was later to be explained by the motion of negatively charged particles.

Hall and Rowland were by no means the only experimental physicists seeking positive confirmation of Maxwell's theory in the 1880s, and rival theories to Maxwell's, as well as reformulations of Maxwell's *Treatise,* were beginning to proliferate both in Great Britain and abroad. Maxwell's physical theory required an ether and denied the existence of an electrical fluid. The theory assumed that matter is different from ether, although perhaps only because of the size of its material molecular vortices. Maxwell's theory supposed that vibrations in molecules of matter produce light, which moves through the ether at a speed *c*. Maxwell's mathematical theory also predicted that waves created by alterations in electric current in the electromagnetic field travel at the speed *c*. How successful was Maxwell's original theory by the end of the century? And how had it changed?

Rivals and Revisions of "Maxwell's Theory"

Hermann von Helmholtz's physics laboratory in Berlin became a major testing-ground for rival theories of electrodynamics in the last third of the nineteenth century. Trained in medicine as well as in physics and physiology, Helmholtz made a formidable reputation for himself with his 1847 publication of the long essay "On the Conservation of Force." Along with that of J. Willard Gibbs (1839–1903) in the United States, Helmholtz's work in thermodynamics helped establish the theory of change in "free energy" as the motor of chemical reaction (see chapter 4). Having been successively professor of physiology at Königsberg, Bonn, and Heidelberg during the period from 1849 to 1871, Helmholtz returned to Berlin as professor of physics in 1871, and he became director of the newly established Physikalisch Technische Reichsanstalt in suburban Charlottenberg in 1887 (see chapter 1).

Familiarizing himself with rival theories of electrostatics and electrodynamics, Helmholtz developed a law of electric potential that could be used to derive different sets of electrodynamic equations by setting a constant $k = 1$, 0, or -1. For Maxwell's equations, $k = 0$; whereas $k = 1$ or -1 for the rival action-at-a-distance formulations.

Helmholtz reiterated his continuing commitment to a hypothesis of electrical fluid in London in 1881, when he was invited to deliver the Faraday Lecture before members of the London Chemical Society. He expressed admiration for Maxwell's theory, which he said he was directing researchers in his laboratory to test. But the most famous part of the lecture became the brief Helmholtz argued in favor of "atoms of electricity" that attach themselves to corpuscles, or molecules, of matter, and he reminded his chemical colleagues of the early theories of Faraday that sought connections between chemical affinity and electrical force.

Helmholtz's laboratory was a center for students and researchers from the United States, Great Britain, and Japan as well as from Germany and central Europe. Rowland and Albert A. Michelson (1852–1931) were among the prominent Americans who studied there as young men. When Heinrich Hertz entered Helmholtz's laboratory at Berlin in 1878, Helmholtz directed him, like others, to work on testing the electrodynamic theories. One of Hertz's first investigations resulted in a demonstration that electricity does not have inertia, as predicted by Weber. Helmholtz suggested that Hertz look for electromagnetic effects that were predicted to result from dielectric displacement currents, but Hertz did not follow up this suggestion while in Berlin, thinking that effects from high-frequency oscillations would be impossible to detect.

In England, Oliver Lodge (1851–1940) was also concerning himself with possible ways to test Maxwell's theory. Lodge, the son of a clay merchant, turned to physics at the age of sixteen after hearing John Tyndall lecture at the Royal Institution. Lodge took an external degree at the University of London, became an assistant and advanced student with George Carey Foster at University College in London, and in 1881 became the first professor of physics at the new University College in Liverpool.

As an experimentalist, not a mathematician, Lodge was a fervent modeler of the ether, an enthusiast for the main points of Maxwell's theory, and a friend and correspondent of two other influential "Maxwellians," George F. Fitzgerald (1851–1901) and Oliver Heaviside (1850–1925).

Fitzgerald was professor of natural and experimental philosophy at Trinity College, London, from 1877 until his death in 1901, and he first met Lodge at the 1878 Dublin meeting of the BAAS. Heaviside, a nephew of Charles Wheatstone, the inventor of the telegraph, worked in submarine telegraphy from 1868 to 1874, publishing articles on telegraphy and electricity in the *English Mechanic* and the *Philosophical Magazine*. After illness in 1874 he worked independently on electrical researches at his family home in Camden Town in North London, and by 1888 he was publishing a series of articles on electromagnetic waves that brought him into correspondence and occasional visits with Fitzgerald and Lodge.

The historian Bruce J. Hunt has characterized these three—Lodge, Fitzgerald, and Heaviside—as an important group of "Maxwellians" with a rather different point of view from Cambridge-educated mathematical physicists William Thomson, Stokes, J. J. Thomson, and indeed Maxwell himself. For one thing, it was Heaviside who reduced Maxwell's equations in the *Treatise* to four equations often called Maxwell's equations. Heaviside dispensed with Maxwell's vector potentials, which still seemed to concede too much to action-at-a-distance, in favor of a vector calculus notation (see figure 1). These equations, which Maxwell had stated in words but not notation in a note of 1868, were also taken up by Hertz in his book on electric waves in 1892. They appeared, perhaps most influentially, in August Föppl's *Einfuhrung in die Maxwell'sche Theorie der Elektricität* in 1894. Albert Einstein was one of the many readers of Föppl.

The "Maxwellians" concerned themselves with the properties of the ether. Fitzgerald demonstrated the fundamental incompatibility

$$\text{div}\,\mathbf{D} = \rho$$

$$\text{div}\,\mathbf{B} = 0$$

$$\mathbf{curl}\,\mathbf{E} = -\,d\mathbf{B}/dt$$

$$\mathbf{curl}\,\mathbf{H} = \mathbf{J} + d\mathbf{D}/dt$$

Figure 1. Maxwell's four equations (modern notation). The mathematical operations *div* and ***curl*** specify directional aspects of forces and fields. The other symbols relate to electric and magnetic force and induction (**D, B, E, H**), to current density (**J**), and to resistivity (ρ).

between Maxwell's ether and the elastic-solid ether that was a mainstay of optical theory since the work of George Green in the 1830s. The Maxwellians ridiculed Stokes's notion of an ether that was "like" wax, pitch, or jelly, and they offered other interpretations. Fitzgerald, for example, suggested in 1898 that lines of electric force were long vortex filaments twisted into corkscrew spirals. A permanent kink in the spiral, with a change in its "handedness" on either side, would correspond to a discrete positive or negative electrical charge.

Lodge, whose conception of the ether included its role as a telepathic medium linking spirits and the living, not only conceived of mechanical and visual models for the ether but sought to test its effects. He, like Fitzgerald, had followed with interest the efforts of Michelson and Edward Morley (1838–1923) in the United States to take up Maxwell's suggestion for finding ether drag.

Albert Michelson carried out his first test for the ether wind in 1881, while he was spending time in Helmholtz's laboratory in Berlin. Michelson was a German-speaker whose family had emigrated to California from Prussia when he was a child. Experiments to test for an ether effect were resumed after Michelson returned to the United States, where he was appointed professor of physics at the Case School of Applied Science in Cleveland in 1883. During the rest of the 1880s he collaborated in these experiments with Edward Morley, a chemist who taught at Cleveland Medical College and Western Reserve College.

Michelson built an interferometer that was meant to detect a difference in the velocity, or wave fronts, of two beams of light traveling at right angles and returning to a common surface. The interferometer, mounted in a stone floating in a basin of mercury in order to minimize extraneous vibrations, could detect an effect of one part in 10 billion.

Michelson and Morley were successful in repeating the experiment by Fizeau that had confirmed Fresnel's theory of a stationary ether, amended by the assumption that transparent bodies have a partial dragging effect on light passing through them. But the failure to detect the earth's motion relative to the ether persisted, and Michelson began to incline to Stokes's view that the earth drags along with it the ether close to its surface.

During the years 1890 to 1893 Lodge was also carrying out experiments with an interferometer and a "whirling machine" to look for ether drag. Lodge later claimed that Fitzgerald's idea that bodies traveling through the ether might change in size or length by just the amount needed to account for Michelson and Morley's null result had grown out of discussions that he had had with Lodge. Fitzgerald, like Lodge and Heaviside, inclined to the view that forces between molecules are related in some way to electricity and to electromagnetic forces, all of these forces being propagated through the ether.

Using Heaviside's recently derived formula for the crowding together of electric and magnetic fields around the middle of a moving charge, Fitzgerald proposed in a letter to the American journal *Science* that matter contracts in the direction of the earth's motion through space. It was through Lodge's publications on the subject in 1892 and 1893 that many physicists, including Henrik Lorentz, learned of Fitzgerald's contraction hypothesis.

It was Fitzgerald, however, who had earlier (in 1879) discouraged Lodge from searching for light effects from the oscillating discharge of a Leyden jar or condenser on the grounds that light is produced by vibrations of atoms and molecules, not by electrical forces. In 1882 Fitzgerald read Lord Rayleigh's *Theory of Sound* (1877) and began to change his mind—coming to think of electromagnetic vibrations, heat, and light as part of one continuous spectrum of frequencies. But the task of detecting high-frequency vibrations that are not visible or heat-producing was still daunting. How could it be done?

While thumbing through an issue of Gustave Wiedemann's *Annalen der Physik* in 1888, Lodge discovered Hertz's paper reporting his recent experiments confirming Maxwell's theory. Its subject matter intrigued him, but his interest was further piqued by the fact that he had met Hertz in Germany in 1881. So Hertz had produced and detected electric waves traveling at the speed of light. How had he done it?

Heinrich Hertz, who originally prepared himself to become an engineer, had finished his physics education under Helmholtz in 1884. After two years as *Privatdozent* at Kiel, he moved to the Technische Hochschule in Karlsruhe, where there was a good laboratory. In 1886 he noticed that the oscillatory discharge of a Leyden jar or induction coil through a wire loop caused sparks to jump a gap in a similar loop a short distance away.

Using Helmholtz's theory of dielectric polarization, Hertz interpreted this effect as an induction phenomenon. He then turned to Maxwell's theory. The finite propagation of electric waves in space or air, independent of a wire, is a central point in Faraday's and

Maxwell's theory. As Hertz began trying to measure the velocity of what he now regarded as waves by reflecting them off the walls of his laboratory, his first results suggested that the velocity of the electric waves was greater than that of light, or perhaps infinitely large (instantaneous). This result favored action at a distance over Maxwell's theory, and it might be a crucial experiment.

Moving his apparatus to his lecture hall, the experimental decision soon turned in favor of Maxwell's theory. Hertz calculated a finite velocity for the electric waves at approximately the speed of light, although he also noticed what he took to be a discrepancy from Maxwell's theory in the electric waves' apparently slower velocity in wires than in air. If this were an accurate observation, then electromagnetic field theory would need some alterations.

Helmholtz announced Hertz's results to the Berlin Physical Society: "Gentlemen! I have to communicate to you today the most important physical discovery of the century." [14] And, indeed, electric waves, or radio waves as they came to be called, proved revolutionary in both their scientific and practical applications. In the next years, most physicists concentrated on reproducing the reflections of electric waves as initially described by Hertz. None replicated his wire guidance experiments, which remain difficult to explain, and Hertz himself turned to other problems.

Hertz later defined Maxwell's "theory" as Maxwell's set of equations, by which he meant the four equations as amended by Heaviside and himself. Hertz did not offer a visualizable description of Maxwell's theory. He knew, of course, that Helmholtz had been working on a theory to eliminate a priori assumptions about matter and force in favor of calculation of the energy of a system defined only by distance between volume elements in characteristic states. Hertz also was sensitive to recent criticisms by Kirchhoff and Ernst Mach (1838–1916) of the foundations of classical physics and of the mechanical world view.

The result was Hertz's publication in 1894 of *The Principles of Mechanics*. Here Hertz rejected the concept of force as a foundation for mechanics, charging that the term *force* lacked clarity. Like Henri Poincaré (1854–1912) and Pierre Duhem, Hertz argued that what is important is not the verification of particular entities or theorems but the verification of a system as a whole. Thus, he wrote, "I know of no shorter or more definite answer than the following: Maxwell's Theory is Maxwell's system of equations."[15]

Radiant Matter and the Etherial Electron

Hertz's studies in electricity in the 1880s included investigations with the electric-discharge tubes known as Geissler tubes in Germany and Crookes tubes in Great Britain. The study of phenomena associated with electrical discharge through gas was another area of investigation that had intrigued Faraday, who had investigated what he called the "glow phenomena" in 1833.

At ordinary room pressure (760 millimeters of mercury = 1 atmosphere of pressure), a spark passes through the air between electrified metallic plates when there is a potential difference of several hundred volts. In the eighteenth century, electrostatic machines could deliver long sparks because voltages were very high, sometimes as high as 30,000 volts. The lower voltage produced by voltaic piles depended on the number of units in the pile. Batteries at well-financed laboratories and institutions, such as the Royal Institution, delivered hundreds or even several thousand volts of electricity. In the last decades of the nineteenth century, the Ruhmkorff coil was invented to produce high potential differences and long sparks. It consisted of a primary coil made of thick wire with a few turns and a secondary coil of thin wire many miles—even hundreds of miles—long.

While Faraday could achieve high voltages in the 1830s, he could not produce the low pressures that were made possible around 1860 by use of the mercury pump, which allowed glass tubes to be evacuated of their gas, bubble-by-bubble, as mercury in a tube sank drop by drop. As air or gas was evacuated from the tube through which an electric discharge passed, a striking sequence of glow phenomena could be observed: first a spark between the positive and negative electrodes placed at right angles or opposite each other; then an elongated, luminous glow with color characteristic of the gas; then the breaking up of the glow into bands or striations with the appearance of a dark space near the anode (the so-called Faraday dark space); then the appearance of phosphorescence near the anode and a new dark space near the cathode first reported by William Crookes; and finally, complete darkness except for a bright, phosphorescent patch at the wall of the tube away from the cathode. The color of the phosphorescence varied with the kind of glass used to construct the tube. The initial part of the sequence occurs at about 20 millimeters of pressure (0.003 atmosphere) and the last phase at a pressure of about 0.000001 atmosphere.

In 1858, just before the introduction of the mercury pump, Julius Plücker (1801–1868) reported that the discharge in the vacuum tube was deflected by a magnet. Using a better vacuum, his student Johann

Hittorf (1824–1914) reported a decade later that an object placed in front of the cathode cast a shadow opposite, indicating that the discharge originated in the cathode.

Eugen Goldstein (1850–1930) coined the term "cathode rays" in 1876, at about the same time that William Crookes began developing a theory that the discharge was matter thrown into a "fourth state" that was not solid, liquid, or gas. Crookes identified this state as one of "radiant matter," harking back to a term Faraday had in 1816 used for a "gradual resignation of properties" in matter, as matter "ascends in the scale of forms."[16]

William Crookes (1832–1919), an experimentalist and the editor of *Chemical News,* launched a series of improvements in the vacuum, coupled with ingenious and astonishing investigations in the discharge tubes. In 1875 he built a "radiometer" by mounting on a pivot in an evacuated glass bulb an arm or axle with vanes at either end. One surface of each vane was painted black, the other white. Crookes explained the continuous rotation of the arm and the vanes as the result of radiation pressure. Some scientists cited Crookes's work as confirmation of the radiation pressure predicted by Maxwell's electromagnetic theory. The explanation, however, turned out to be more complex, having to do with the gas still remaining in the not perfectly empty vacuum bulb.

While continuing to experiment with his radiometer, Crookes introduced similar techniques to study the "cathode rays," demonstrating, for example, that a paddlewheel mounted on a glass track within the electric discharge tube moves toward the anode under the "impact" of the rays on its mica vanes. He confirmed that a shadow was cast away from the cathode if an aluminum cross was placed in the tube and that the phosphorescent light at the end of the tube could be deflected by a magnet. He demonstrated that the rays could be focused by a concave or convex cathode. He concluded that the cathode rays were not radiations at all, but particles of "radiant matter" independent of the nature of the gas residue in the discharge tube and negatively charged, since repelled by the cathode:

> In studying this fourth state of matter we seem at length to have within our grasp . . . the little indivisible particles which with good warrant are supposed to constitute the physical basis of the universe. . . . We have actually touched the border land where Matter and Force seem to merge into one another.[17]

It is not surprising that some sober-minded German theoretical physicists rejected the spiritualist-sounding conclusions of Crookes. (Indeed, Crookes was a believer in spiritualist phenomena.) Hertz was among those who set out to disprove Crookes's "fourth state of matter."

In 1883 Hertz placed the discharge tube between electrostatically charged plates and reported that there was no deflection of the radiation. If Crookes were right about the negative charge of the rays, a deflection of the phosphorescent patch toward the positively charged plate would be expected. It was possible, of course, that enough gas was left in the tube such that polarized or ionized gas particles were attracted to the charged plates, insulating them from the rays, but this did not seem feasible to Hertz.

Hertz's student Philipp Lenard (1862–1947) demonstrated in 1892 that the cathode rays penetrate a thin aluminum window soldered into the discharge tube, which also seemed proof that they could not be particles of radiant matter. Further, the rays blackened photographic plates, like light. The Germans seemed to agree that cathode rays are an electromagnetic wave disturbance in the ether, and this is where the matter stood in the early 1890s.

As we have already seen, rival conceptions of the ether, its relation to matter, and the existence or nonexistence of electrical particles existing between lines of magnetic and electric force were numerous and much contested by the end of the nineteenth century. It can be misleading to describe differences along strictly national lines, but, on the whole, it can be said that German and Continental physicists tended to characterize the ether in terms that were more abstract and less realist than those used by British physicists (and chemists). Two concluding and contrasting examples are the theories of electromagnetism developed by Larmor and Lorentz in the late 1880s and early 1890s.

Joseph Larmor was born and educated in Belfast before going to Saint Johns College, Cambridge, where he finished as senior wrangler in the mathematical tripos in 1880, edging out J. J. Thomson. Like Thomson (to whom we shall return in chapter 6), Larmor in the 1880s interested himself in the properties of a fluid ether that was so endowed with latent rotational elasticity that it would allow the propagation of transverse waves but was sufficiently inertial in its properties that it could support the rotation of vortex rings.

After working with the idea that positive and negative electrification might be spread homogeneously over the core of an ether vortex, Larmor in 1893 broke the electrification apart into pointlike charged nuclei of rotational strain, which he called "monads." The negatively charged cathode-rays, he suggested, might be streams of these monads.

In correspondence with George Fitzgerald in 1894, Larmor conceived of a better name than "monad"—namely, "electron," a term that Fitzgerald's uncle George Johnstone Stoney (1826–1911) had used for ionic or electrical charges. Larmor's *Aether and Matter,* which appeared in 1900, derived properties of matter, electromagnetism, and light from the mechanical ether, which was taken to be the primary material of the universe. So pervasive and persuasive was this approach that a few chemists, too, could be found speculating about the material ether at the end of the nineteenth century. Adolf von Baeyer, for example, proposed that the density of ether might be greater inside than outside the benzene ring.

Whereas Larmor derived most physical phenomena from the ether, Henrik Lorentz took the opposite tack by deriving the laws of ordinary mechanics, including inertial mass, from the electromagnetic condition of the ether. Lorentz, who held the first chair of theoretical physics at the University of Leiden from 1878 to 1923, was recognized by the turn of the century as one of the leading physicists on the continent, largely because of his work in electrodynamics. He was influential both because of his acumen as a physicist and his cosmopolitan charm as a colleague.

Like Larmor, Lorentz sought unifying principles through a physics of the ether and by 1900 he was confident that physical and chemical phenomena and perhaps even gravitational effects might be encompassed in an electron theory rooted in the ether. Emil Wiechert (1861–1928) in Germany and Paul Langevin (1872–1946) in France were working along similar lines.

For Lorentz, however, the ether had no mechanical connection with matter; rather, in the spirit of Maxwell, he drew a sharp distinction between ether and matter. The interaction between the two occurs through tiny, inertial, rigid bodies found in molecules and carrying positive or negative electrical charge. In 1895 Lorentz called these bodies "ions" and later, after 1899, "electrons." Thus electric flow is the flow of electrons. This hypothesis is, of course, contrary to Maxwell's argument that electricity is not material. For Lorentz, the motions of electrons create an electromagnetic field, the seat of which lies in the stationary ether.

Lorentz also concluded that electrons, and thus molecular matter, are affected by the speed of their motion, becoming shorter at speeds approaching the speed of light. This contraction would explain the failure of Michelson and Morley to detect the motion of matter through the ether. Lorentz soon realized that Fitzgerald had earlier come to the same conclusion about the relationship between the length and the speed of an electron as it moves through the ether.

A confirmation of Lorentz's theory came from experimental work by his younger Dutch protégé Pieter Zeeman (1865–1943) in 1896. Following up on Faraday's researches, Zeeman tried to repeat one of Faraday's last investigations, which had failed. In 1862 Faraday had looked for the effect of a magnetic field on the spectral lines emitted by sodium vapor. Maxwell, who believed that spectral lines were caused by the vibrations or oscillations of molecules independent from the ether and electromagnetic phenomena, had denied the existence of the effect sought by Faraday. Zeeman, who had at his disposal a diffraction grating far superior in resolving power to Faraday's prisms, observed a slight broadening of the spectral lines in the magnetic field. He found that he could produce spectral line triplets or doublets by varying the orientation of the direction of observation and the magnetic field.

This Zeeman effect gave Lorentz and Zeeman a tool to calculate the ratio of electron charge to electron mass (e/m) on the hypothesis that the light that is the spectral lines is emitted by the electrons of charge e and mass m moving about in regular periodic fashion within the atom. The charge for e was negative and the value of the ratio was approximately a thousand times larger than expected. Lorentz took this result to mean that the electron (or ion), if it carried a standard unit of charge, was very small.

The convergence of spectroscopic phenomena with electrical-discharge phenomena was to provide a fertile meeting ground for theories of matter, electricity, and light in the 1890s and the early twentieth century. Lorentz's results seemed to confirm the notion of "atoms of electricity" that Helmholtz had advanced in his Faraday Lecture in order to kindle "anew the interest of chemists in the electrochemical part of their science."[18]

As we shall see in later chapters, chemists in the nineteenth and early twentieth century were indeed to find their interest in electricity rekindled, first through Svante Arrhenius's ionic theory and then through J. J. Thomson's electron theory. By the 1920s, electrons were to provide explanatory mechanisms for chemical problems that were left unresolved in the theories of both structural organic chemistry and thermodynamical physical chemistry.

The Continental physicist Albert Einstein disposed of the ether in a relativistic physical theory that Faraday might well have liked (see chapter 7), no matter how much Larmor, Lodge, and William Thomson in fact disliked it. The direction in which the electrodynamics of the electron developed during the 1920s and 1930s was foreshadowed by Fitzgerald and Lorentz in the early 1890s, but hardly foreseen.

J. Norman Lockyer (1836–1920), the astronomer, spectroscopist, and editor of *Nature,* provided a striking metaphor for matter, electricity, and light that was to become outmoded by the 1920s:

> We believe that each molecular vibration disturbs the ether; that spectra are thus begotten; each wavelength of light resulting from a molecular tremor of corresponding wavelength. The molecule is, in fact, the sender, the ether the wire, and the eye the receiving instrument, in this new telegraphy.[19]

This vision of the natural world was a powerful one. It was a worldview that greatly complicated Newtonian principles or rejected them outright. The challenge of thinking about electricity and magnetism during the course of the nineteenth century brought into relief the difference in scientific explanation between realist visualization and abstract mathematics, even as it highlighted some differences in national customs of building scientific theories.

4

Thermodynamics, Thermochemistry, and the Science of Energy

A New Challenge to Newton's Mechanics

If the electromagnetic view of nature diverged radically from the Newtonian tradition of action-at-a-distance, it did not break with the Newtonian conception of "force." Michael Faraday's experiments led him to imagine lines or tubes of force running between electrically charged centers or between magnetic poles. Similarly, in their mathematical development of Faraday's theory, William Thomson and J. C. Maxwell concerned themselves with what they called the electromotive force.

Yet, Thomson, like others, also gradually began to develop a new concept that rivaled and, in some important respects, replaced "force." This was the concept of "energy." The development of a precise definition of energy, along with statements of the conservation of energy as a law of nature, was a slowly evolving process that became a pronounced scientific movement around the 1850s. This process included the gradual realization that heat cannot be a material substance. By century's end, some scientists were beginning to argue that only energy, not matter or force, was the primary reality of the universe.

Thomson provided one of the first important definitions of the term *energy*, which was used in the nineteenth century by Thomas Young in his discussion of the seventeenth-century mechanics of *vis-*

viva, literally, the "living force" produced by bodies in motion and described geometrically in the past by Simon Stevin, Christiaan Huygens, Gottfried Leibniz, and others.

In 1854 Thomson coined the term *thermo-dynamics* and by the end of the nineteenth century thermodynamics was claimed to be a new science unifying physics and chemistry, with many scientists hoping to derive both Newtonian mechanics and Continental electrodynamics from the fundamental principles of thermodynamics, or "energetics," as it was also called. By 1914, Newton's three laws (inertia, force in relation to mass and acceleration, and action and reaction) were complemented by three laws of thermodynamics: the conservation of energy, the increase of entropy, and Nernst's heat theorem.

Like Newton's laws, these thermodynamic laws were often said to have been "discovered" through sheer genius. But in contrast to Newton's laws, there was some controversy about *who* "discovered" the three laws of thermodynamics. Some students of Walther Nernst (1864–1941) claimed that in his lectures in Berlin in the early 1900s Nernst said that the first law of thermodynamics had been discovered by three scientists (students presumed them to be J. R. Mayer, James Joule, and Hermann von Helmholtz), the second by two scientists (Sadi Carnot and Rudolf Clausius), and, as for the third, "Well, this I have just done by myself."[1]

The science of energy, or thermodynamics, has often been said by historians, particularly social historians, to have originated in the study of practical problems associated with the technology of the steam engine. Both Joule and Carnot were familiar with these kinds of industrial and engineering problems. The nineteenth-century origins and subsequent history of "energetics" have also been associated by historians, particularly intellectual historians, with early and late Romanticism, with German *Naturphilosophie,* and with an anti-mechanistic or anti-Enlightenment mood that exalted living force, *Kraft,* power, even "will" over inert, brute matter and over materialistic reductionism. There is some validity to each of these historical accounts.

Another important and not unrelated set of origins for thermodynamics, however, lies in the experimental investigations and precise measurements that sought to establish physical equivalents of "force," "power," or "activity" that would be comparable to the chemical equivalents for matter and similarly quantitative in nature. As we saw in chapter 3, Faraday established a precise equivalence between weights of elements that combine through chemical affinity and weights of elements that are deposited at electrodes by electrical force.

Faraday also measured the quantities of heat associated with the activity of the battery (its force), the decomposition of electrolytes (the work accomplished), and the weights of elementary substances deposited at the cell's electrodes (the electrochemical equivalents). This was a subject to which James Joule turned in his private laboratory in Manchester in the 1840s. His aim was to establish a precise figure for the "mechanical equivalent of heat" that would have a significance comparable to chemists' figures for equivalent weights and to Faraday's figure for the electrochemical equivalent.

In this chapter, we begin with Joule's work and its integration into British mathematical natural philosophy by Thomson, Peter Tait, and William Rankine. We will look at Continental traditions of thermochemistry (including Julius Thomsen, Marcellin Berthelot, and Pierre Duhem) and the transformation of experimental thermochemistry by mathematical thermodynamics (Helmholtz, Clausius, Maxwell, Ludwig Boltzmann, and J. H. van't Hoff), including the work of the American mathematical physicist J. Willard Gibbs.

In 1887 Svante Arrhenius, Wilhelm Ostwald, and van't Hoff announced the foundation of what they called a new discipline of physical chemistry that would unify the practice of chemistry and physics, most specifically bringing together thermochemistry and thermodynamics, although still failing to solve the problem of the origin of chemical affinity.

Preoccupation with energetics led to physical and metaphysical interpretations that seemed to undermine the Newtonian framework of mechanism and determinism at the very time that Jean Perrin, among others, demonstrated experimentally that the corpuscular and molecular assumptions at the basis of the kinetic theory of matter were correct. At this same time, Max Planck and Albert Einstein offered arguments for the discrete and atomistic nature of energy. Thermodynamics, probably more radically than electromagnetism, created the conviction among many scientists that the laws describing natural phenomena did not have a set of consistent and harmonious foundational principles.

Mechanical Invention, Mathematical Physics, and the Conservation of Energy

Like his mentor John Dalton, James Prescott Joule (1818–1889) was a largely self-educated man. In contrast to Dalton, Joule was born into the comfortable circumstances of a wealthy English entrepreneurial family. His father owned a brewery at Salford, near Manchester, and the young Joule's early interests in machines, chemistry, and electricity could easily be indulged in experiments at the brewery and at a laboratory in his home. Both Joule and his brother studied chemistry and natural philosophy with Dalton, who taught them the pleasures of exact measurement.

Like Dalton and most other British enthusiasts for the sciences, Joule believed that natural laws exist and that they express the mind of God. In the tradition of Francis Bacon, he wrote in a notebook that "the first object of natural science is to elevate humanity in the scale of creatures, and the second is to promote well-being."[2] One means of promoting well-being was to improve the machinery in Manchester breweries and factories. Toward this end, the teenage Joule pondered the notion of a perpetual-motion machine or, at the very least, an improvement of the electrical motor, a device that was invented when Joule was fourteen years old by John Sturgeon, a lecturer at the Royal Military College in Surrey.

In 1837 Joule constructed a motor powered by a chemical battery and found it could raise fifteen pounds a foot high in a minute. From then on, Joule continued to calculate the work that could be done by employing a chemical battery, electromagnetism, falling bodies, or other means. His interests and his calculations began to focus on the heat that always seemed to be associated with work, including the heat generated by the passage of water through narrow tubes or by compression of air with a hand pump or by the churning of water with paddle wheels. He arrived at the law for the heat generated through resistance in an electric circuit (the i^2r law), and he calculated precise values for the mechanical equivalent of heat.

As Joule carried out his experiments and calculations, he was influenced by contemporary writings of Dalton, Humphry Davy, and Faraday. Joule assumed that matter is made of atoms, that chemical affinities of atoms are in some way associated with electricity, and that heat is generated by the forces required for the combination and separation of atoms. An early paper assumed that atoms are surrounded by atmospheres of electricity and that the changing motion—that is, rotational speeds—of these electrical clouds accounts for heat effects.

For Joule, then, heat was not a substance but a state of motion. While others had earlier ascribed to the view that heat is motion, Davy among them, this was not a common view in scientific journals in the 1840s. When Joule ran into difficulty on more than one occasion in getting his papers published by the Royal Society of London, he tended to see himself as very much out of the mainstream: "I could imagine those gentlemen in London sitting around a table and saying to each other, `What good can come out of a town where they dine in the middle of the day.'"[3]

Joule's results did not go unnoticed, however, and when he presented a paper to the chemistry section of the BAAS in Oxford in 1847, Thomson showed up and asked questions. Thomson was strongly interested in Joule's demonstration that heat is produced by the friction of fluids in motion, but he was unnerved by Joule's claims that heat is a form of action and power, like chemical action or mechanical action, rather than a form of matter.

The tradition of treating heat as a material fluid and the assumption of the conservation of heat had resulted in strikingly successful mathematical results, particularly in the French engineer Robert Clapeyron's applications of Carnot's heat cycle as described in Carnot's *Refléxions sur la puissance motrice du feu [Reflections on the Motive Power of Fire* (1824)]. When he met Joule, Thomson was a twenty-three-year-old graduate of the Cambridge mathematical tripos and a new professor of natural philosophy at the University of Glasgow. He had mastered French mathematical physics firsthand in Paris as well as in his assiduous study of books and journals.

In contrast, Joule was treating heat nonalgebraically in the tradition of *vis-viva.* In an 1847 lecture at the reading room of Saint Ann's Church in Manchester, Joule had stated, as he now repeated at Oxford,

> You see, therefore, that living force may be converted into heat, and that heat may be converted into living force, or its equivalent attraction through space. All three, therefore—namely, heat, living force, and attraction through space (to which I might also add light, were it consistent with the scope of the present lecture)—are mutually convertible into one another. In these conversions nothing is lost. . . . We can therefore express the equivalency in definite language applicable at all times and under all circumstances. Thus the attraction of 817 lb. through the space of one foot is equivalent to, and convertible into, the living force possessed by a body of the same weight of 817 lb. when moving with the velocity of eight feet per second, and this living force is again convertible into the quantity of heat which can

increase the temperature of one pound of water by one degree Fahrenheit.[4]

For Joule, who had studied under Dalton and who took chemistry seriously, there not only existed equivalents of matter that can be expressed in numbers corresponding to relative combining weights, but there also existed equivalents of action that can be expressed in interconvertible units of *vis-viva*.

Joule was by no means the only one thinking along these lines. Julius Robert von Mayer (1814–1878), a German physician, concluded that less chemical oxidation of the blood is required to produce body heat in the tropics than in cooler climes. He expressed his confidence in a principle of conservation of *Kraft* meaning "force" or "power."

Hermann von Helmholtz, as a young German medical doctor who had studied physiology in the Berlin laboratory of Johannes Müller, reasoned similarly in his essay "On the Conservation of Force [*Kraft*]" in 1847. The essay, which began by stating the impossibility of perpetual motion and the primacy of Newtonian central forces, laid out definitions of what Helmholtz called tensional force (*Spannkräfte*) and living force (*lebendige Kraft*), defined in terms of potential (mgh) and actual (mv^2) effect. The essay specifically applied these definitions to traditional mechanical problems such as gravitation; to heat effects, including the heat formed by chemical processes; and to electrical batteries, magnetism, and electromagnetism, as well as to living organisms.

But it was William Thomson, struggling with Joule's experimental results and Clapeyron's mathematical expressions, who first clearly defined "energy" by way of contrast to force, on the one hand, and to heat, on the other. In a footnote to a paper he published in 1849, Thomson expressed his perplexity over the problem of reconciling Carnot and Clapeyron with Joule on the dissipation of heat in a solid body. If Joule were correct, "When 'thermal agency' is thus spent in conducting heat through a solid, what becomes of the mechanical effect which it might produce? Nothing can be lost in the operations of nature—no *energy* can be destroyed. What effect then is produced in place of the mechanical effect which is lost?" [Emphasis added.][5]

By 1851 Thomson realized that the mechanical effect was not lost but was transformed into the motion of the particles making up the solid body. This conversion of heat into motion, in turn, could produce work, for example the work done by a gas as it expands against a piston in an idealized Carnot engine. In collaboration, Joule and

Thomson found that the temperature falls for a gas that is allowed to expand without doing work, as predicted if internal work is necessary to overcome attractions between the gas molecules. (This discovery became the basis for the early refrigeration industry.)

In 1851, in "On the Dynamical Theory of Heat," Thomson wrote that, for any material body, "it is convenient to choose a certain state as standard for the body under consideration, and to use the unqualified term, *mechanical energy,* with reference to the standard state." He distinguished "statical" (later, "potential") from "dynamical" (later, "kinetic") energy, on the basis of origin. By the time he gave a lecture at the BAAS annual meeting in Liverpool in 1854, Thomson was claiming that Joule's discovery of the conversion of heat into work constituted the greatest reform in physical science "since the days of Newton."[6]

In addition, by the mid-1850s Thomson had been corresponding and collaborating with William Rankine (1820–1872), an engineer by training. Coining the term energetics, Rankine argued in 1855 that the new science of energy had no requirement for a particular theory of matter or force but was based in experimental observations free from hypothesis. The notion of the independence of energetics and thermodynamics from hypothesis became a leading theme by which the science of energy came frequently to be distinguished from the older Newtonian science of action-at-a-distance forces and the newer electromagnetic science of ether-and-matter fields.

In 1867 Thomson and his longtime friend Peter Guthrie Tait published a volume entitled the *Treatise on Natural Philosophy.* It decisively made the case for natural philosophy as a science of energy by substituting a dynamics of energy for the Newtonian dynamics of force. Fourier-type analysis could be used to describe heat flow; a Hamiltonian function represented potential and kinetic energy; and energy, or work, was the product of the motion of points or bodies. The *Treatise* was written as a university textbook to replace the atom-and-force framework of Newton, Laplace, and Poisson with a dynamical theory of matter-in-motion in the tradition of continuum mechanics.

From Thermochemistry to the Second Law of Thermodynamics

While in Paris in 1845, the twenty-one-year-old Thomson had obtained an introduction to Henri-Victor Regnault, who was thought by many of his colleagues, in France and abroad, to be the best physicist in France. Regnault taught both chemistry and physics during his long career, and his laboratory at the Collège de France was well equipped for the precision measurements which were his passion. While his reputation was enormous in his lifetime, his renown diminished afterwards because he was not associated with any major theory or discovery.

Accepted as an assistant in Regnault's laboratory, the young Thomson spent many days working from eight o'clock in the morning until five or six in the evening. He learned how to work air pumps, to seal glass tubes, and, as he said later, to emulate "a faultless technique, a love of precision in all things, and the highest virtue of the experimenter—patience."[7]

Regnault was especially respected for his experimental researches on heat, and, when Thomson arrived, Regnault was taking up the problem of the compressibility of gases, a subject that was influential in Thomson's subsequent collaborations with Joule. As Regnault investigated the fit between precise experimental results and Boyle's law of gases, he discovered that all gases have slightly different coefficients of expansion. He worked out improved values for atomic weights based on measurements of specific heat in the law of Dulong and Petit.

Regnault was a member of a generation of men whose interests lay in the relations between the physical and chemical properties of matter. Robert Bunsen, Hermann Kopp, J. L. Meyer, and H. H. Landolt (1831–1910) all introduced new or improved instruments, including the spectroscope, calorimeter, polarimeter, and vacuum pump, into the chemical and physical laboratory. Like Dmitry Mendeleyev, they calculated physical quantities such as molecular or atomic volumes (the ratio of atomic weight to density), specific heats, melting and boiling points, refractive indices, and gas expansion coefficients, and they classified their results in relation to atomic weight and to chemical combining properties.

The emphasis on quantitative investigations of heat relations soon revolutionized traditional approaches to the old problem of chemical affinity, shifting emphasis from force to heat and then to energy.

First Julius Thomsen (1826–1909) in Copenhagen and then Marcellin Berthelot in Paris inaugurated systematic studies based on the assumption that the chemical force, or affinity, holding parts of a molecule together must be directly proportional to the heat evolved during chemical reaction. In other words, they assumed that the heat that evolved in this "exothermic" reaction represented the work to overcome chemical forces of affinity.

Berthelot, who became during his lifetime the single most powerful scientific administrator in France, self-consciously set out early in his career to reform affinity studies. He switched the focus from acid-base reactions, which occur fairly quickly, to organic reactions, which take considerably more time. He developed a new instrument, the bomb calorimeter, an instrument that became so thoroughly identified with his laboratory that Thomsen, involved in priority disputes with Berthelot from the 1870s on, refused to use it.

As it turned out, both Thomsen (in 1854) and Berthelot (in 1864) were wrong in their argument that the chemical reactions that actually occur are those in which the greatest amount of heat is produced, an argument that Berthelot called "the principle of maximum work." Neither was Berthelot correct in designating the active mass in chemical reaction as the quantity of chemical substance. In contrast, a proposal by two protégés of Thomsen, the Norwegians Cato Maximilian Guldberg (1836–1902) and Peter Waage (1833–1900) proved more fruitful in the long run. They defined "active mass" in terms of *molecular* concentration.

More immediately, within Berthelot's own Parisian laboratory, a gifted student criticized Berthelot's approach in a doctoral thesis that was turned down in 1884 at the University of Paris. The student was Pierre Duhem, who finally got his doctoral degree by writing a different thesis, while nonetheless finding himself "banished" to the provinces rather than being granted one of the teaching positions in Paris that Berthelot controlled.

What Duhem had done was to bring to bear on Berthelot's experimental studies the new mathematical theories of heat developed outside France in the work of Clausius, Gibbs, and Helmholtz. This work fundamentally altered the science of heat and the calculation of chemical reactions by the end of the nineteenth century by supplementing thermochemistry with two laws of thermodynamics.

In 1850 Clausius proposed clearly and decisively that there must be two fundamental laws in thermodynamics, not one, if the results of Carnot and Clapeyron were to be consistent with Joule's experimental results favoring a motion, or dynamical, theory of heat. To the accepted principle that heat and work are interconvertible (the principle of conservation of energy), Clausius added a second principle: that the maximum work accomplished in a reversible process, for example the expansion of a gas, depends on the quantity of heat transferred, but that in order to do work, this transfer can occur only from a higher to lower temperature or from hotter to colder bodies.

Thus there is an important limitation on the freedom of interconvertibility, or "conversion of forces," as it was often put.

Clausius then expanded the scope of the two principles of thermodynamics from gases (the steam engine) to electrical discharge and to thermoelectricity. In a paper of 1854 he developed a mathematical expression for the "equivalence" value of transformations between heat and work, whatever the source of the mechanical work accomplished. In a completely ideal cycle of reversible processes, all transformations would cancel each other out, so that their summation could be represented by a mathematical function:

$$\int \frac{dQ}{T} = 0$$

where Q is quantity of heat and T is absolute temperature (the temperature in relation to an ideal state of zero heat).

Therefore, it is the quantity dQ/T, not Q, which is conserved under the circumstances of an ideal heat engine. For irreversible processes in isolated systems, using his second principle, Clausius took the difference $(T_2 - T_1)$ to be a positive quantity where T_2 is the higher temperature. Thus, the sum of transformations for irreversible processes is positive.

In 1865 Clausius coined the word *entropy* (from the Greek for "transformation") for the mathematical transformation function, and he restated the two fundamental laws of thermodynamics as "The energy of the universe is constant" and "The entropy of the universe tends to a maximum." Thus for infinitesimal reversible processes, where S = "entropy,"

$$dS = \frac{dQ}{T}$$

and for spontaneously occurring processes,

$$dS > \frac{dQ}{T}$$

By this time Clausius had defined the term *disgregation* to mean a measure of the arrangement of molecules in a body, and he identified increasing entropy with increasing molecular disorder or positive disgregation.

In a widely translated and reprinted lecture, Clausius generalized these results from physics to metaphysics:

> One hears it frequently said that all the world has a circular course, . . . that the same states are reproduced constantly and that the state of the world remains invariable when one considers the world at large and in a general manner. The first principle of the mechanical theory of heat perhaps was considered a confirmation of this principle. . . . Now we have found a natural law which permits concluding in a certain way that in the universe all does not have a circular course, but that modifications take place in a determined sense, and tend thus toward a limiting state.[8]

Clausius was not alone in this interpretation. Using the language of the "dissipation" of energy, Thomson wrote in his 1851 paper "On the Dynamical Theory of Heat" that heat flow in natural processes is directional and irreversible. This he took to have philosophical and religious significance. And, in a book coauthored with Balfour Stewart in 1875, which carried the dramatic title *The Unseen Universe,* Peter Guthrie Tait argued that energy passes back and forth between seen and unseen universes and that our physical world will ultimately dissipate into an eternal world beyond our boundaries.

Very different was the approach of J. Willard Gibbs, who, in teaching a course in potential theory at Yale University, made use of the work of Clausius as well as of a short note by F. Massieu (1869) deriving entropy, internal energy, pressure, and volume from a single function. In the early 1870s Gibbs submitted several papers to the Connecticut Academy of Sciences, among them a many-part paper "On the Equilibrium of Heterogenous Substances," which was published from 1875 to 1878.

Gibbs's starting point was the principle that a system, whether chemical, thermal, or mechanical, is in a state of equilibrium when the entropy of the system has reached a maximum value and the internal energy is at a minimum value. By the late 1880s Gibbs's application of his fundamental equations to different "phases" of bodies existing in equilibrium with each other attracted the interest of the Dutch professor of physics Johannes D. van der Waals (1837–1923) and his younger protégé Hendrik Roozeboom (1854–1907), who found it useful in understanding the equilibrium of sulfur dioxide and water. Following Ostwald's publication of a German translation of Gibbs's memoir in the early 1880s, Gibbs's work came into the hands of Duhem and the mainstream of European scientific literature.

By then Helmholtz, too, had turned his attention to refining the mathematical treatment of thermodynamics. Based on studies of the galvanic battery and electrochemistry, Helmholtz argued that only part of the energy of chemical reaction in the battery is expressed as heat and that the rest is transformed or capable of being transformed into mechanical work. This latter part he called "free energy," treating it as a kind of potential energy analogous to the potential function in mechanics, much as Gibbs had done earlier. The free energy is defined as

$$F = U - JTS$$

where J is the mechanical equivalent of heat, U is the total internal energy, T is absolute temperature, and S is entropy. If temperature is held constant, then there is a decrease in free energy equal to the maximum amount of work achieved when the reaction is carried out reversibly.

Chemical affinity in the chemical battery and in electrolysis, then, is to be identified with free energy or with work, not with heat. As historian Helge Kragh has noted, it was only with Helmholtz's version of the energy/entropy/heat function that its implications for the problem of chemical affinity were clearly understood. It was Helmholtz's version that van't Hoff and Nernst further explored. The importance for chemistry of Helmholtz's work was publicly acknowledged in 1892 when the German Chemical Society elected Helmholtz to honorary membership.

In the language of chemistry, the term *affinity* gave way to the term *energy,* just as, in the language of physics or natural philosophy, the term *force* also ceded rhetorical and intellectual space to *energy.* Physical chemistry became a kind of chemistry focused on energy relations in chemical reactions, thus shelving the problem of the causes of elective affinities of individual atoms for study by a later generation.

Reversibility, Irreversibility, and New Interpretations in Physical Chemistry

What linked together late-nineteenth-century physicists' and chemists' interests in thermodynamics and "energetics" was an experimental methodology and a mathematical approach that focused on reversibility and irreversibility in natural phenomena. This common interest was a novel circumstance because, in the early nineteenth century, "physical" phenomena had often been distinguished from "chemical" phenomena on the basis of the contrasting categories of "continuous versus discontinuous" and "reversible vs. irreversible" phenomena.

The phenomena that early-nineteenth-century natural philosophers, or physicists, studied by mathematical means included moving bodies, light, heat, and, later, electromagnetism. Their mathematical methods were those of the algebraic calculus, founded in infinitesimal analysis with attention to differential increments of time. The fundamental proposition embedded in mathematical mechanics, made explicit, for example, by Laplace, was the reversibility and predictability, in principle, of the motion of all corpuscles or points in the universe.

In contrast, chemists in the early nineteenth century established laws of definite and constant proportions for chemical combinations of the elements, as well as sharply defined atomic or equivalent weights and molecular weights. The chemical processes most easily studied were irreversible in character. Once a salt had precipitated or a gas had evolved, the reaction was finished, and the results could be analyzed meticulously and at leisure, with no concern for velocity or time.

But by the mid-nineteenth century, these distinctions were beginning to break down. The study of gases was leading in physics to the notion of irreversibility, as made mathematically explicit in Clausius's, Thomson's, and others' statement of a "second law" for the theory of heat in 1850 and 1851. And in chemistry, systematic studies in organic synthesis were leading to the recognition of the importance of time, as well as variations in mass, temperature, and pressure, for the many instances of chemical equilibrium in which there was no clear end to the reaction but rather an ongoing redistribution and reversibility of reactants and products in solution.

Within the tradition of thermochemistry, Marcellin Berthelot, in collaboration with Léon Saint-Gilles (1832–1863), worked on these kinds of problems in studies, for example, of the reaction between ethyl alcohol and acetic acid. Henri-Étienne Sainte-Claire Deville (1818–1881), while the director of the chemical laboratory at the École Normale Supérieure in Paris, studied the dissociation of gases at high temperatures, with attention to reversibility and irreversibility. By the 1880s the work of the German physicist Ludwig F. Wilhelmy (1812–1864), a lecturer in Heidelberg who had studied with

Regnault in Paris, was also becoming better known. Wilhelmy had investigated the rates of hydrolysis of sucrose to "invert sugar" (a mixture of glucose and fructose). These rates could be calculated by observing with a polarimeter the change in rotation of polarized light from right to left as the sucrose was converted. Wilhelmy found that the rate of inversion depended on the concentration of sucrose. For Berthelot, Sainte-Claire Deville, and Wilhelmy, precise measures of physical conditions of chemical reactions linked chemistry and physics together.

More perhaps than any other figure in the late nineteenth century, the Dutch-born scientist Jacobus H. van't Hoff (1852–1911) successfully combined interests and achievements in chemistry and physics, concluding his career in Berlin as professor at the Academy of Sciences and at the University of Berlin. He was awarded the first Nobel Prize for Chemistry in 1901 for his work on rates of reaction, chemical equilibrium, and osmotic pressure, but he is equally well known to chemists and chemistry students for proposing in 1874, independently of Joseph-Achille Le Bel, the "carbon tetrahedron": the arrangement in three-dimensional space toward the four corners of a tetrahedron of the four chemical valence-bonds of the carbon atom.

In addition to enrolling in a mathematics and physics curriculum in the Netherlands, van't Hoff studied under Kekulé in Bonn and Wurtz in Paris. Van't Hoff taught physics briefly at the State Veterinary School of Utrecht, for which he was ridiculed by the Leipzig chemist Hermann Kolbe in an infamous attack:

> A Dr. J. H. van't Hoff, of the Veterinary School of Utrecht, finds, it seems, no taste for exact chemical research. He has considered it more convenient to mount Pegasus (apparently on loan from the Veterinary School) and to proclaim in his `La chimie dans l'espace' how, during his bold flight to the top of the chemical Parnassus, the atoms appeared to him to be arranged in cosmic space. . . . It is typical of these uncritical and anti-critical times that two virtually unknown chemists, one of them at a veterinary school and the other at an agricultural institute, pursue and attempt to answer the deepest problems of chemistry . . . with an assurance and an impudence which literally astounds the true scientist.[9]

About the same time that van't Hoff launched stereochemistry, he also concerned himself with the motion of atoms in their molecular space. During 1878 to 1881 he published a two-volume treatise on organic chemistry in which he developed a mathematical treatment of chemical affinity using gravitational analogies for the forces holding atoms together in molecules. Unsatisfied with this theory of chemical dynamics, he employed a new approach in publications in 1884 and 1885, which he wrote in French since he thought that French chemists were more open to the use of physical methods and mathematical formulas than were German chemists.

Drawing on the work Helmholtz, August Horstmann, and others, van't Hoff newly demonstrated that the heat of reaction is not a direct measure of affinity. If it were, reversible reactions would be impossible because the products associated with maximum evolution of heat would be those of maximum stability. He introduced the now-familiar double arrow sign for denoting equilibrium, noted the importance of the concentrations of reactants, described the condition for dynamic or mobile equilibrium as one of equality of forward-and-backward reaction rates, and derived an expression for the equilibrium constant. Among his experimental examples of reversibility and equilibrium were the dissociation of nitrogen dioxide studied by Sainte-Claire Deville,

$$N_2O_4 \Leftrightarrow 2NO_2$$

and the esterification of acetic acid investigated by Berthelot and Saint-Gilles.

At about this same time, Henry-Louis Le Châtelier (1850–1936), a professor at the École des Mines in Paris, stated the general principle that when a change is imposed on a system in dynamic equilibrium, the system will respond in a way that tends to reduce its effect. There were few departures from these basic formulations in the next decades (indeed, in the next century). Even so, because they found conditions of constant pressure more amenable to their work than constant volume, chemists (particularly in the United States) began using Gibbs' expression for free energy or maximum work rather than Helmholtz's and van't Hoff's expression.

Van't Hoff came to realize the widespread applicability of principles rooted in the theory of heat and in the study of reversible equilibrium. A discussion with his Amsterdam colleague, the botanist Hugo de Vries (1848–1945), about osmotic pressure in plants led van't Hoff to recognize that osmotic pressure, a phenomenon not much studied, could provide a measure of the affinity of a salt solution or a sugar solution for water by measuring the pressure required to stop the flow of water into the solution. Van't Hoff studied the applicability of the ideal gas law $PV = RT$ to osmotic pressure as well as to dilute solutions. In particular, he used data from François-Marie Raoult's freezing-point depressions and boiling-point elevations of solvents in dilute solutions as correction factors in his equation $PV = iRT$, where i was adjusted to fit experimental data.

Van't Hoff's work was quickly taken up by Svante Arrhenius (1859–1927) in Uppsala and by Wilhelm Ostwald in Riga, Latvia. In 1887 Arrhenius published a new theory, based on his doctoral dissertation. He argued the rather surprising idea that, at infinite dilution, all molecules of electrolytes in water break up into charged ions, even if there is no battery current running through the solution. François-Marie Raoult (1830–1901) was on the verge of this statement in his calculations of the molecular lowering of the freezing points for water, which Arrhenius first noted when reading van't Hoff's discussion of it. But Raoult had not made clear the view that dissociated "radicals" in solution are charged particles, or "ions"; instead, he had used his data to calculate molecular weights.

In the late 1870s, the ambitious and industrious young Ostwald fastened on the aim of constructing a table that would systematize values of affinity in the way that Mendeleyev and Meyer had systematized atomic weights. But by the time Ostwald took up a new chair of physical chemistry at Leipzig in 1887, energy, not affinity, had become the focus of his attention. "As long as we sought to measure chemical 'forces,'" he wrote in an article in 1887, "the theory of affinity made no progress."[10] Energy was now the gleaming sword that would cut the Gordian knot.

By 1887, Ostwald, Arrhenius, and van't Hoff were in contact with each other, and in that year they, as coeditors, launched the first journal devoted solely to physical chemistry, the *Zeitschrift für physikalische Chemie.* Arrhenius came to Leipzig to work briefly with Ostwald before returning to teach in Stockholm, and Ostwald hired the young Nernst as his assistant that year.

Ostwald became an efficient and effective proselytizer for what he called the new discipline of physical chemistry, which was to be a general chemistry of fundamental laws and principles. To this end, he published more than five hundred scientific papers, forty-five books, and five thousand reviews during his career, in addition to overseeing the translation into German of significant papers including those of Gibbs, Guldberg and Waage, and van't Hoff in the still ongoing *Klassiker der exakten Wissenschaften.*

By 1900 the experimental work and theories of the "Ionists," as the Ostwald group often was called, were still controversial, particularly among organic chemists. In addition, a program of "energetics" proselytized by Ostwald and backed in various degrees by other physicists and physical chemists became similarly controversial. Ostwald aimed to explain all of chemistry, physics, and, indeed, biology with recourse only to experiences of energy mediated by measurements of heat. The hypotheses of the atom and the molecule would disappear. By the end of the nineteenth century, controversy was joined over the rapidly developing kinetic theory of gases and the statistical interpretation of the second law of thermodynamics.

A Dynamical Theory of Molecules

In 1857 Rudolf Clausius made explicit the molecular assumptions that underlay his earlier papers on heat. An impetus for this work was a paper recently published by August Krönig, the editor of the journal *Fortschritte der Physik.* Clausius's view, which belonged to a mathematical tradition going back to the Swiss mathematician Daniel Bernoulli in the eighteenth century and revived by Krönig, was that gas molecules do not vibrate around equilibrium positions but move uniformly in straight lines until they collide with other molecules or with the walls of their container. Bernoulli had arrived at a mathematical expression for this theorem, which can be expressed

$$p = \frac{1}{3}\left(\frac{nmv^2}{s^3}\right)$$

where p is pressure, n is the number of molecules, m is the mass of a single molecule, v is the mean velocity of the molecules, and s is the length of a side of a cubic container.

Interest in the dynamical theory of heat, in combination with his experimental studies of gases, had led James Joule to publish a paper in 1848 in which he calculated the velocity of a molecule of hydrogen, based on the reasoning of Bernoulli. In this paper, Joule combined his result (6,225 feet per second at 60°) with reasoning about the mechanical equivalent of heat in order to calculate the specific heat of hydrogen.

Clausius in turn developed a theoretical formulation by treating gas particles as elastic spheres and assuming that forces between particles can be ignored in developing a general law. He also treated what Joule had called *vis-viva* as energy that varies with temperature, leading to the general formula

$$pV = \frac{1}{3nm\bar{v}^2}$$

where v^2 is the molecule's mean square velocity.

Knowing the density at a given pressure, Clausius deduced that the average speed of gas molecules must be several hundred meters per second. He also assumed that the exchange of energy between rotational and translational motions of gas molecules would equilibrate to a constant ratio, and he worked out an equation predicting the ratio for a gas molecule of specific heat at constant pressure to specific heat at constant volume (c_p/c_v). Allowing six degrees of freedom for a diatomic molecule, for example, his prediction was

$$\frac{c_p}{c_v} = \frac{(n+2)}{n}$$

where, for example, with $n = 6$, then $c_p/c_v = 4/3 = 1.33$.

The following year Clausius published a second paper, in which he responded to criticism that molecules could not possibly be traveling as fast as he had calculated because pungent odors and drifting smoke take a considerable time—minutes rather than seconds—to fill a room. This very argument had been made against Bernoulli's theory by Joseph Priestley in the eighteenth century.

Clausius countered the criticism by proposing that gas molecules undergo large numbers of collisions as they move about at high velocities, so that their effective speeds are considerably smaller than their real velocities. Applying statistical reasoning to the problem, as he had done earlier in a treatment of the blue color of the sky, he calculated the probability of a molecule's covering a given distance without collision as being very small. Clausius's treatment assumed this "mean free path" *l* to be related to the "sphere of action" or effective diameter *s* of the molecule by the equation

$$\frac{1}{l} = \frac{4}{3\pi s^2 N}$$

where *N* is the molecules' number density.

Clausius's paper attracted the attention of J. C. Maxwell, whose 1855 Adams Prize essay at Cambridge had tackled the problem of investigating the motions and stability of Saturn's rings, which Maxwell treated as collections of millions of small bodies. Now Maxwell focused on the behavior of gas molecules as a possible instance or application of a mathematical theory of the collisions of particles.

Maxwell based this new theory on his earlier reading of Laplace and George Boole as well as on John Herschel's essay in the 1850 *Edinburgh Review* reviewing Adolphe Quételet's *Theory of Probability as Applied to the Moral and Social Sciences.* Statistical techniques and calculation of probabilities had a long history in the calculation of outcomes for games of chance, deaths, and crime, but they had not been applied to the kinds of physical objects that are the subject of natural laws. Unlike statistical laws, natural laws were claimed to be absolutely certain.

In Maxwell's "Illustrations of the Dynamical Theory of Gases" he calculated a bell-shaped distribution curve for the average velocity of gas molecules, showing that, while an average velocity for gas particles at a given temperature could be assumed, individual molecules actually move both faster and slower than the average. The curve took the form of the familiar Laplacian or Gaussian curve for distribution of error, as used by Quételet for distributing measures of behavior to arrive at the *homme moyen* (the average, or normal, man).

Just as Quételet's work led to discussions of the implications of "moral statistics" for free will and determinism in human conduct, so Maxwell felt compelled to distinguish what he called "moral certainty" from "absolute certainty." Statistical laws describing the behavior of systems of individual particles or bodies result in calculations of averages or means that carry only highly probable validity. In contrast, dynamical laws treating the motions of individual particles result in predictions with absolute validity. The law of conservation of energy was a law of absolute certainty, but the law of increasing entropy and irreversibility, Maxwell soon began to think, was not so.

Maxwell's paper of 1860 derived an expression demonstrating that equal volumes of gases contain equal numbers of particles, or molecules, as Avogadro's law required. Maxwell also derived expressions relating the viscosity of a gas to its density as well as ones predicting the molecules' mean free path and average velocity. During the early 1860s, Maxwell and his wife, Katherine Dewar

Maxwell, collaborated to determine experimentally the mean free path of oxygen molecules at 0°C. They found the value to be 5.6×10^{-6} cm, a number in good agreement with results from gas diffusion experiments. They also found that the viscosity of gases is a function of pressure but that it is not independent of temperature. This result contradicted the hypothesis of elastic-sphere molecules, leading Maxwell in 1867 to propose a theory of molecules as centers of force that repel one another according to an inverse fifth-power law.

A problem that came to beset the statistical, or kinetic, theory was the assumption, shared by Clausius, Maxwell, and Ludwig Boltzmann, of the "equipartition" of the energy of a system of molecules among the internal mechanical motions or degrees of freedom of motion of individual molecules. Experimental information that refuted their assumption continued to accumulate after 1860. Most gases were found to have a c_p/c_v ratio of 1.41, not 1.33. Then, in 1875, August Kundt and Emil Warburg found experimentally that the ratio for mercury vapor was 1.67, that is, the value expected if this gas were composed of molecules with no internal degrees of freedom, so that all heat energy results in translation of the molecules.

In 1877 Boltzmann proposed that diatomic molecules might have only five degrees of freedom, reasoning that collisions of molecules that are ellipsoids of revolution would not change the amount of rotation around the axis. His theoretical value then was in agreement with experiment (7/5 = 1.41).

Maxwell, however, was less sanguine than Boltzmann (who did not change his opinion) about the equipartition dilemma, saying,

> We learn from the spectroscope that a molecule can execute vibrations of a constant period. It cannot therefore be a mere material point, but a system capable of changing its form. . . . Some molecules can execute a great many different kinds of vibrations. They must, therefore, be systems of a very considerable degree of complexity, having far more than six variables. . . . Every additional variable, therefore, increases the specific heat, whether reckoned at constant pressure or at constant volume. I have now put before you what I consider to be the greatest difficulty yet encountered by the molecular theory.[11]

Altogether abandoning the model of elastic spheres in his widely read article "Atom" in the ninth edition of the *Encyclopaedia Britan-*

nica, Maxwell suggested that a vortex model of the atom, treated mathematically, was now the great challenge for the theoretical or mathematical physicist. This mathematical model would require the use of some one or more absolute constants that were doubtless essential in the construction of all material molecules:

> From a comparison of the dimension of the buildings of the Egyptians with those of the Greeks, it appears that they have a common measure. Hence, if no ancient author had recorded the fact that the two nations employed the same cubit as a standard of length, we might prove it from the buildings themselves. . . . Each molecule . . . throughout the universe bears impressed on it the stamp of a metric system as distinctly as does the metre of the Archives at Paris or the double royal cubit of the Temple of Karnac.[12]

But if equipartition and its meaning for the model of an atom was a great problem for Maxwell, it was not the only one. Following Maxwell's 1867 paper introducing centers of force into the theory of gases, Boltzmann worked out from mechanical principles a statistical distribution law of molecular motions deriving a theoretical function *H,* which can only remain zero or decrease as the system comes to equilibrium. Since the mathematical function *H* corresponds to the negative entropy function *S,* Boltzmann provided a basis in molecular mechanics for the second law of thermodynamics. His derivation presumed, however, the classical Newtonian and Laplacian mechanical principle of the reversibility of all velocities of all particles in motion.

In the meantime, Maxwell began to think of the second law as distinctly peculiar, when interpreted in terms of the dynamical theory of molecules. He wrote to Peter Tait in 1867 saying that he had begun to imagine an interloper, a "finite being" (whom William Thomson soon termed a "demon") who could observe the motions and velocities of individual gas molecules. In a letter to Lord Rayleigh in December in 1870, Maxwell explained:

> Put such a gas into a vessel with two compartments and make a small hole in the AB wall about the right size to let one molecule through. Provide a lid or stopper for this hole and appoint a doorkeeper very intelligent and exceedingly quick. . . . Whenever he sees a molecule of great velocity coming against the door from A into B he is to let it through,

> but if the molecule happens to be going slow, he is to keep the door shut. He is also to let slow molecules pass from B to A but not fast ones. . . . In this way the temperature of B may be raised and that of A lowered without any expenditure of work. . . . The 2nd law of thermodynamics has the same degree of truth as the statement that if you throw a tumblerful of water into the sea, you cannot get the same tumblerful of water out again.[13]

Just as Maxwell was pondering the problem of irreversibility, Boltzmann's Viennese colleague Josef Loschmidt (1821–1895), whose work in molecular physics included the first good estimate of the diameter of a molecule, questioned Boltzmann about what Loschmidt called the "irreversibility paradox." How could the experimentally established irreversible process of the increase of entropy be reconciled with the reversible and time-invariant processes of the molecular-kinetic theory?

In 1877 Boltzmann responded to Loschmidt with a paper defining the second law of thermodynamics as a statistical law based in the tendency of molecular systems to reach their most probable thermodynamic state, namely a state of thermal equilibrium. Boltzmann's formula

$$S = k \log W$$

where S is entropy, k is a constant, and W is the probability of a state, appears on a monument dedicated to Boltzmann's memory in the city of Vienna. However, Boltzmann's confidence that mechanical explanation remained the best explanation was coming under severe siege in German and Austrian scientific circles by the 1890s.

A Sense of Crisis: Anxieties about Mechanics and Thermodynamics

By the early 1890s both molecular dynamics and the kinetic-molecular theory had more than a few ardent critics. In a lecture at the Johns Hopkins University in 1884, William Thomson warned that the equipartition problem was one of two "dark clouds" threatening the bright horizon of progress in natural philosophy. (The other cloud was the inability to detect experimental effects from the ether.)

As theoretical physicists and physical chemists were rethinking the foundations and interconnections of their disciplines at the century's end, opinions were voiced that the time had come to end the reign of classical mechanics as the basis of physical science.

Thermodynamics, some argued, was a science embodying a new and higher stage of exact and positive knowledge, free of metaphysical or fanciful presuppositions. Even the young J. J. Thomson, who later became well known as an advocate of mechanical hypotheses, praised thermodynamics as a science based "only" on empirical quantities measured in the physical laboratory. Pierre Duhem sought to reestablish mechanical laws on the basis of thermodynamics. Wilhelm Ostwald stunned members of the London Chemical Society in 1904 with a lecture arguing that the fundamental laws of the chemical atom (the laws of constant proportions, multiple proportions, and combining weights) could be deduced from thermodynamics alone. He argued that all forms of matter and force in the natural world are the manifestations of one real but immaterial substance, which is energy.

Debates and controveries about the merits of mechanics and thermodynamics were common in scientific meetings and in the pages of scientific journals at the turn of the century. What, scientists and other intellectuals asked, were the roles that molecular theories, force theories, and energy theories would play in the future?

Recalling his students days in Vienna in the early 1900s, Phillipp Frank (1884–1966) wrote that two characteristic beliefs of nineteenth-century science seemed to be breaking down: first, the belief that all phenomena can be reduced to the laws of mechanics, and second, the belief that science will eventually reveal the truth about the universe. Postmodernists of the late twentieth century hardly invented antipositivist critiques.

The flight from mechanical physics was based not simply in empirical and logical anomalies but also in a pervasive and gloomy spirit of fin-de-siècle malaise. Ostwald's anti-atomism, like the anti-mechanistic views of the Austrian physicist and philosopher Ernst Mach, registered a resistance to and rejection of the earlier optimism of many scientists about the applicability of mechanics to problems of perception, physiology, and mind. Mach took the view that an understanding of the world's reality cannot be constructed on physics alone but must also be based in biology. While not ascribing to Ostwald's "energetics," Mach shared Ostwald's view that the laws of thermodynamics have greater use in biology than do the laws of classical mechanics.

In Paris, one of the best-attended lecture halls at the turn of the century was that of Collège de France philosopher Henri Bergson

(1859–1941). Many students, colleagues, and members of the city's cultural elite reveled in Bergson's spirited and poetic monologues extolling intuition and antirationalism. He argued the priority of biological time ("lived time"), as experienced and perceived by individual beings, over artificial representations of space-time ("mechanical time").

Identifying "energy" with force or will *(énergie-volonté)*, and interpreting the second law of thermodynamics as a law of history and freedom, Bergson and many of his contemporaries revived the values of early nineteenth-century romanticism and idealism against the principles of Enlightenment mechanism and materialism. "Energy" was vitality and freedom; "matter" was inertness and determinism.

Not all interpreters of the metaphysical meanings of the laws of thermodynamics took Bergson's hopeful view. Many feared that there might be dreadful cosmic implications in the second law of thermodynamics. The law of irreversibility might doom the universe to losing all motive force as useful energy was gradually transformed into a steady state of uniform heat, resulting in a "heat death" of all matter and life. H. G. Wells's short novel *The Time Machine* (1895) was a terrifying fantasy both because of the future postindustrial world it depicted and because of the lifeless sun that barely illuminated that future world.

Scientists' statements that the law is an ideal law for precisely defined conditions hardly assuaged the apprehensions of those inclined to pessimism.

Kinetic Theory and the Quantum Theory by 1911

Among those who took a different point of view were members of a circle of young Parisian physicists and chemists that included Jean-Baptiste Perrin (1870–1942), Paul Langevin (1872–1946), and Marie and Pierre Curie. When they gathered at Sunday dinners, their talk was of the Dreyfus affair, the works of Wagner, the philosophical opinions waxing in Paris, and, of course, their own laboratory work. Committed to the usefulness and indeed the truth of molecular-kinetic theory, Perrin lectured in the early 1900s at the Sorbonne on its recent triumphs.

The diameter of molecules was estimated at 10^{-7} to 10^{-8} centimeters, and better and better estimates of Avogadro's number were being made. Johannes van der Waals's law of corresponding states demonstrated that the gas laws could be reliably applied to liquids.

The work of François-Marie Raoult showed the success of these assumptions in calculating improved values of molecular and atomic weights based on freezing-point depression and boiling-point elevation of solutions.

Perrin's early experimental work on X rays and cathode rays as well as the Curies' investigations of the extraordinary phenomenon of "Becquerel rays" (radioactivity) were providing further proofs of the existence of discontinuous or discrete ions and molecules. In 1905 Perrin began studying a phenomenon that was to prove especially fruitful in confirming the explanatory power of the molecular kinetic theory: Brownian motion.

In the 1830s the Scottish botanist Robert Brown (1773–1858) systematically studied the spontaneous motions of pollen granules in water under a microscope. Like the French botanist Adolphe Brongniart, Brown initially speculated that the motion might be caused by the life force within the pollen. Further experiments, however, revealed similar motions for suspended microscopic particles of metals, rocks, and even a fragment of the Sphinx.

By the time Perrin turned to this topic in the early 1900s, the many explanations of the phenomenon included surface tension, capillarity, temperature effects, electrical action, and, more recently, the kinetic motion of the visible microscopic particles as they are bombarded by the invisible molecules of the liquid in which they are suspended.

With Langevin advising him on calculations, Perrin estimated the mean energy of yellow vegetable latex granules from observations of their motions, but his value came out some 100,000 times smaller than that required by the kinetic theory. The new ultramicroscope invented by Henry F. W. Siedentopf (1872–1940) and Richard Zsigmondy (1865–1929) in 1903, however, made possible observations at the limits of 5×10^{-3} microns. Perrin soon recognized that the apparent mean velocities of observed particles varied in size and direction, depending on the length of the time of observation, without tending toward a limit. Thus, the observed motions were not velocities at all but rather displacements that were the visible results of the many collisions and changes of direction of the microscopic particles moving at faster-than-visible speed.

From 1905 through 1912 Perrin organized a laboratory program of investigations on colloidal suspensions of latex, mastic, and resin particles. He studied three principal phenomena. These included the vertical distribution of colloid particles after they reach equilibrium (this corresponds to the Laplacian distribution of particles in air under the influence of gravity), the translational displacement of particles, and the rotation of particles. Perrin's experimental procedures were complicated and precise, requiring separation (by centrifugation) of particles with identical radius, establishing their density, marking them with small faults or occlusions for studies of their rotation, and taking thousands and thousands of photographs of particles to be counted under well-defined conditions.

One of Perrin's principal aims was the calculation of the value *N* in equations from the kinetic theory of gases following his measurements of the density of the granules, their mass, the number of granules per unit volume of water, the pressure exerted by a granule, and the temperature. His very first series of experimental investigations, focusing on the distribution of particles under the influence of gravity, turned out to be an effect predicted by Albert Einstein in 1905.

Einstein's paper was brought to Perrin's attention by Langevin after Perrin had begun his experimental work on the subject. In the paper, "On the Movement of Particles Suspended in Fluids at Rest, with the Aid of the Molecular-Kinetic Theory of Heat," Einstein used considerations based in entropy, free energy, and osmotic pressure to calculate the diffusion and the displacements of small suspended particles. The displacements, he predicted, would be proportional to the square root of the time:

$$\lambda_x = \sqrt{t}\sqrt{\frac{RT}{N}\cdot\frac{1}{3\pi kP}}$$

Einstein realized in constructing his theory that the proposed movements might be identical with the so-called Brownian motion: "If the movement discussed here can actually be observed, . . . an exact determination of actual atomic dimensions is then possible. On the other hand, [if] the prediction of this movement proved to be incorrect, a weighty argument would be provided against the molecular-kinetic conception of heat."[14]

Einstein and Perrin were soon in correspondence about Perrin's work, and they were to meet in Brussels in 1911 at the first of the international physics conferences sponsored by the Belgian industrialist Ernest Solvay (1838–1922). At that meeting, Perrin displayed values of *N* calculated from a variety of different phenomena relating to Brownian motion (vertical distribution, diffusion, displacements, rotation) in comparison to values of *N* calculated from the viscosity of gases, the blue of the sky, radioactivity, and black-body radiation. The convergence of these values gave empirical certitude to the kinetic-molecular theory. Perrin had finally "demonstrated the existence of atoms and the possibility of counting them."[15] Ostwald gave up his opposition to atoms and admitted that Perrin's experimental studies of Brownian motion, as well as the independent studies by Theodor Svedberg (1884–1971), had convinced him that atoms and molecules were real.

CONSEIL DE PHYSIQUE SOLVAY
BRUXELLES 1911
HOTEL METROPOLE

GOLDSCHMIDT PLANCK RUBENS LINDEMANN HASENOHRL
NERNST BRILLOUIN SOMMERFELD DE BROGLIE HOSTELET
SOLVAY KNUDSEN HERZEN JEANS RUTHERFORD
LORENTZ WARBURG WIEN EINSTEIN LANGEVIN
PERRIN MME CURIE POINCARE KAMERLINGH ONNES

Members of the first Solvay Council of Physics, convened in Brussels in 1911. Of these members, Max Planck, Heinrich Rubens, Frederick Lindemann, Walther Nernst, Arnold Sommerfeld, Maurice de Broglie, Ernest Solvay, James Jeans, Ernest Rutherford, Henrik Lorentz, Wilhelm Wien, Albert Einstein, Paul Langevin, Jean Perrin, Marie Curie, Henri Poincaré, and Heike Kamerlingh Onnes are mentioned in the text. *Courtesy of the late Francis Perrin.*

The 1911 Solvay Conference marked a programmatic attempt, particularly on the part of Henrik Lorentz and Walther Nernst, to set out an agenda for theoretical physics and theoretical chemistry. This agenda aimed at the integration of the disparate and overlapping understandings of molecular mechanics, electromagnetism, and thermodynamics.

Perrin's presentation at the meeting seemed conclusive on the reality of the discrete nature of matter. But another theory of discontinuity had begun to trouble and excite the physicists and physical chemists at the meeting. What were they to make of their colleague Planck's hypothesis of the discrete nature of energy?

Max Planck (1858–1947) had been educated and inspired within the thermodynamic tradition. His early work was part of the new physical chemistry, as well as of theoretical physics. He studied at the University of Berlin under Kirchhoff and Helmholtz before completing a dissertation at the University of Munich in 1879 on thermodynamics and reversibility. Without knowing it, he repeated some of the work of Gibbs. Called to Berlin in 1889, Planck became a junior colleague to Helmholtz and in 1893 to van't Hoff.

During the 1880s and 1890s Planck mostly worked on the task of writing a textbook in thermodynamics that would reformulate the fundamentals of this science in a fashion that reconciled energetics and mechanics, taking into account the theory of ions and the applications to solutions of the kinetic theory of gases but without retreating, as he saw it, into the physical assumptions of atomism and reductionism. On this, he and van't Hoff seem to have agreed. A first edition of Planck's *Foundations of Thermodynamics* appeared in 1893 and a second one in 1897. He was widely respected as a theoretical physicist concerned with physical chemistry.

As a professor in Berlin, Planck included among his wide circle of colleagues experimental physicists who were doing research at the Physikalisch Technische Reichsanstalt (PTR) in Charlottenburg, a suburb of Berlin, some of whom also taught at the city's university or Technische-Hochschule. During the 1890s and early 1900s a principal preoccupation in the Optics Laboratory at the PTR was the study of the relationship between the luminous intensity of a body and its temperature and elementary nature.

As Otto Lummer (1860–1925) explained in a popular lecture in 1897 at the Berlin Polytechnische Gesellschaft, there was practical value to this work. Physical laws on emissivity were needed so that the lighting industry could have an emission standard for the minimum temperature at which light would become visible, as well as formulas for the light-radiation properties of different materials (carbon, metals, metal oxides).

During 1893 and 1894 Wilhelm Wien (1864–1928) worked out a law showing that in the normal emission spectrum of a black body (a body that is a perfect absorber and emitter of all frequencies of light), the product of the wavelength and the temperature is a constant. He treated black-body radiation as an instance of stable heat equilibrium that could be understood in terms of the law of entropy. Experimental apparatus that was built and refined in the next years for studying this radiation included, first, a porcelain furnace with a hole in it and, then, in 1898, a thermocouple placed inside a completely closed platinum cylinder whose blackened interior walls could be brought by electrical means to thermal equilibrium.

Planck worked to derive the current radiation formulas from first principles. Many of these formulas were *ad hoc* equations that had been devised to deal with the better data that the newer laboratory oven supplied. Unfortunately, the improved data were inconsistent with Wien's (and Planck's) earlier equations at high temperatures and at long wavelengths. In the fall of 1900 Planck came up with a new treatment that fit the data very well, based in an electromagnetic version of a theorem of Ludwig Boltzmann's (the "H-theorem"), in which Boltzmann's function approached a stationary value with time.

In Boltzmann's treatment, the total energy continuum for electromagnetic resonators was subdivided into P elements of size ϵ, so that $P\epsilon = Nu$, where N = the number of resonators and u = the energy of the resonators. In contrast to Boltzmann, Planck made the size of the energy elements ϵ, ϵ', ϵ'', etc. fixed and proportional to frequency. In setting up his final function, Planck considered only one set of resonators with frequency f and he omitted the computation of a maximum value, as carried out by Boltzmann in his use of the analogous function. A constant h appears in the energy relationship

$$\epsilon = \mathrm{hf}$$

Planck began to call h the "quantum of action" or the "element of action," modeled on the classical principle of least action.

Initially, Planck's formulation was not regarded as especially advantageous until debates in the journal *Nature* in 1905 made clear the unacceptability of a rival formula from Lord Rayleigh and James Jeans (1877–1946). This formula led to what was dramatically christened an "ultraviolet catastrophe": the prediction of dangerous radiation from a low-temperature oven. As a result, physicists' attention came back to Planck's treatment.

Paul Ehrenfest (1880–1933) and then Einstein took up Planck's law of 1900, with Einstein concluding that Planck's formulation imposed a necessary restriction on the energy of any resonator: that it was either zero or an integral multiple of an "energy quantum" hf. Planck, as late as 1910, still expressed reservations about adopting a restriction that meant that energy resonators could not act continuously, saying that "The introduction of the quantum of action h into the theory should be done as conservatively as possible, i.e., alterations should only be made that have shown themselves to be absolutely necessary."[16]

But by the time of the Solvay meeting in 1911, classical thermodynamics was being transformed by this new theory of the discontinuity of energy, even as classical physics and chemistry were being transformed by the theory of the discontinuity of matter. Both theories were statistical and probabilistic in nature, and both rightly were regarded as revolutionary. All this was made clear in another paper published by Einstein in 1905, this one entitled "On a Heuristic Point of View about Light," which was a topic of discussion at the 1911 meeting.

In this paper Einstein used methods of statistical mechanics similar to those developed by Gibbs to demonstrate that entropy of radiation takes a mathematical form typical of a gas. If light, as one form of electromagnetic radiation, is treated as a collection of particles with energy proportional to the frequency of light, then, Einstein correctly suggested, there is a good explanation of the dependence of electron emission from metals on the frequency of incident light rather than on the intensity of light. Further, as now expressed by Einstein, the value of the constant h in Planck's equation,

$$E = hf$$

can be calculated directly from experimental information on the photoelectric effect, because

$$E = hf - P$$

where *P* is the work done in the ejection of an electron from a metal. The physical implication of this formulation is that the energy of light is not distributed evenly over the whole wave front, but it is concentrated in discrete small regions or elements. In response to Einstein's work, Planck himself had begun to use the phrase "quantum of energy" (*hf*) by 1909, drawing an analogy from the discontinuity of energy to that of matter (the atom) and electricity (the charge *e*).

The physicists and physical chemists invited to the 1911 Solvay Conference constituted a small and elite group. Among the twenty-two present were the German scientists Planck, Nernst, Heinrich Rubens, Arnold Sommerfeld, and Wien; the British scientists Jeans, Ernest Rutherford, and Frederick Alexander Lindemann; the French scientists Langevin, Perrin, Marie Curie, and Henri Poincaré; and the Dutch scientists Lorentz and Heike Kamerlingh Onnes. Einstein, who was present, had just moved from Zurich to Prague.

Nernst, who was among those strongly impressed by the somewhat bohemian-styled Einstein, opened the Solvay meeting by comparing its importance to the congress of chemists held at Karlsruhe in 1860, which had met to try to iron out differences and to agree on fundamentals about a coherent system of atomic weights. A molecular conception of matter had again been precipitating problems, he noted, insofar as the recent elaboration of the kinetic and statistical theory had led to formulations conflicting with experimental results from the study of radiation and specific heats, particularly at low temperatures.

Nernst's own heat theorem (the "third law of thermodynamics") for low temperatures, describing how entropy becomes zero at absolute zero, was one of the new theories that promised successfully to connect fundamental principles and laboratory results. Nernst's work correlated thermodynamics and thermochemistry by providing the means for calculating values of specific heats and predicting the likelihood of many chemical reactions through considerations of equilibria, although still not addressing the reaction mechanism for the individual molecule. As developed through statistical mechanics and the restrictive requirements of the quantum theory, discrepancies between laboratory reality and the kinetic theory were disappearing.

But, asked Nernst, at what cost to the theoretical edifice of classical mechanics? By the meeting's end, it was clear that while the quantum theory was a useful device—*Hilfsmittel* as Nernst and Einstein termed it—the quantum hypothesis was contributing to a crisis of confidence and a premonition of a turning point among physical scientists. As Sommerfeld wrote to Lorentz on the eve of the Solvay meeting, "I too am now a convert to the theory of relativity. . . . However, I am sufficiently old-fashioned to resist temporarily the Einsteinian conception of the light quanta."[17]

5

Theory and Practice in Organic Chemistry: Biological Modes of Thought in a Physical Science

Virtues to Valences

For most of the nineteenth century, the methods and problems of the chemistry of hydrocarbons defined the mainstream of teaching and research among those who called themselves chemists. Experimental papers published in the weekly *Berichte* of the German Chemical Society, founded in 1862, were principally concerned with results in organic chemistry, mostly defined in terms of what came to be called classical structural chemistry oriented toward the analysis and synthesis of natural products.

Whereas by the end of the nineteenth century a distinction came to be made between "inorganic" and "organic" chemistry, no such subfields existed in the seventeenth and eighteenth centuries. Chemical substances were identified according to their origins. Following old traditions in natural philosophy and natural history, these sources were the animal, vegetable, and mineral kingdoms. Members of the Paris Academy of Sciences set out as early as 1670 to establish the relationship between the "virtues" of plants and their chemical compositions as well as to analyze animal materials such as blood and lymph for their essential quality-bearing chemical principles.

As described by historian Frederic L. Holmes, these seventeenth- and eighteenth-century chemists debated whether methods of extraction or distillation should be used in their investigations, with the Berlin court apothecary Caspar Neumann seeking to isolate and reconstitute crystalline substances as they "naturally exist in the subject,"[1] not as simple bodies.

While the tradition of Newtonian natural philosophy—namely, chemical explanation in terms of elementary corpuscles, masses, and affinities—remained a crucial component of chemical thinking during the nineteenth century, the tradition of natural history was equally strong, if not dominant. Practitioners of plant chemistry, animal chemistry, agricultural chemistry, and physiological chemistry investigated constituents and products derived from natural organisms. These products, once analyzed into their very simplest parts, were all found to have the common elements of hydrogen and carbon, at least.

The role of natural history in nineteenth-century chemistry is a leading theme in what follows in this chapter. So is the development of the molecular "structure" theory within the natural history tradition, a theory that explains the activities of chemical molecules in the biological language of form and function rather than in the mechanical language of matter, motion, and force. One of the most significant theoretical developments of nineteenth-century chemistry, the chemical valence theory, arose from organic theories of type and structure, not from the mechanical force theory.

The twentieth-century explanation of valence theory by the electron theory is discussed in chapter 6, in light of the renewed interest among many twentieth-century chemists in physical theories, especially electrical theories as the foundation of principles of chemical affinity. As twentieth-century chemists attempted to synthesize larger and larger molecules, the instruments of physics became more and more common in chemical laboratories. As we will see at the conclusion of this chapter, investigations of proteins were important in triggering new debates among natural-products chemists about the usefulness and accuracy of physical theories and physical methods in solving chemical problems.

The Natural History Tradition and the Theory of Molecular Types

A famous chemistry story centers on an evening at the Tuileries palace in Paris (once part of the Louvre but destroyed in 1871), when guests found themselves coughing and crying as the palace's splendid white candles began to emit noxious, acidic fumes. The chemist Jean-Baptiste-André Dumas (see chapter 2) was asked by his father-in-law, the Sorbonne mineralogist and geologist Alexandre Brongniart, to look into the problem. Dumas concluded that the fumes were hydrogen chloride gas, meaning that some of the chlorine used to whiten the candles must have combined with the candle wax. Dumas found that the chemical literature mentioned similar reactions noted by other chemists, including Michael Faraday, who had reported the reaction of chlorine with ethylene.

Dumas soon published a paper, which did not receive much notice, suggesting that "chlorine possesses the singular power of separating the hydrogen from certain bodies and of replacing it atom for atom." This reaction, he claimed, was an example of a "law of substitution" that, along with the "law of addition," must be a common rule of chemical composition and decomposition by which atoms (or equivalents) and organic "roots" or "radicals," consisting of persistent groupings of atoms, might enter into combinations.

In 1837, Dumas found himself presiding over the jury for the doctoral thesis of Auguste Laurent, his former assistant, who was now working on the problem of the action of chlorine, bromine, and nitric acid on coal-tar products such as napthalene. Laurent's thesis directly took issue with Jöns Jacob Berzelius's explanatory scheme for chemical combination by electrochemical forces. Laurent offered in its place a geometrical interpretation of chemical constitution.

Laurent's "nucleus" theory for hydrocarbons depicted a rectangular prism representing a radical, or "nucleus," of eight carbon atoms joined with twelve hydrogen atoms (C^8H^{12}). In this prism, an atom of chlorine or half an atom of oxygen might take the place of a hydrogen atom. This would result in a substituted, stable, "derived nucleus," which could then further combine with atoms (see figure 2). Alternatively, if a volume or unit of chlorine or oxygen were *added* to the face of the prism, pyramidal addition products, not substitution products, were obtained. In all of this, the position of the atom, not its nature or charge, was the primary determinant of chemical behavior.

In 1838 Dumas published a paper on the chlorination of organic substances in which he presented a theory of what he called chemical *types*. He described the production of trichloracetic acid by the action of chlorine on acetic acid:

> [C]hlorinated vinegar is still an acid, like ordinary vinegar; its acid power has not changed. It saturates the same quantity of base as before. . . . So here is a new organic acid in

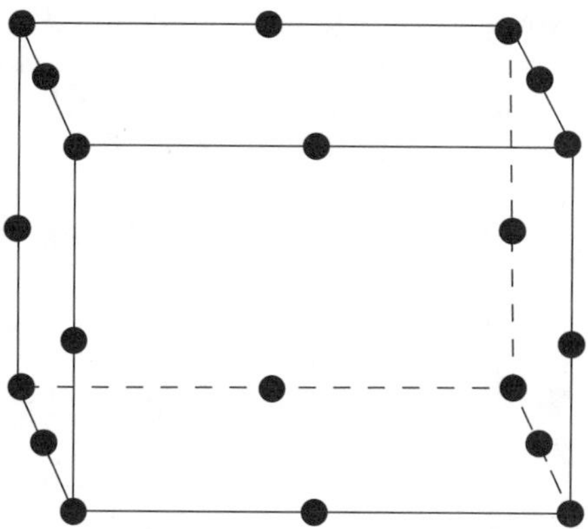

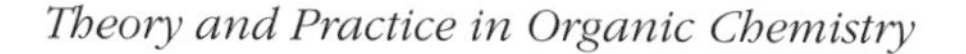

Figure 2. The hydrocarbon nucleus according to Auguste Laurent. Twelve hydrogen and eight carbon atoms each are represented by a dark circle. Addition may take place on the two smaller faces of the rectangular prism, while substitution takes place at the noncorner circles where hydrogen is present.

> which a very considerable quantity of chlorine has entered, and which exhibits none of the reactions of chlorine, in which the hydrogen has disappeared, replaced by chlorine, but which experiences by this remarkable substitution only a gradual change in its physical properties.[2]

Dumas's argument stressed the malleability of the chemical "type," in this case, the organic acid, which maintains its fundamental properties even when new and very different elements enter into its constitution.

This theory met considerable ridicule from some of Dumas's colleagues across the Rhine. Berzelius and Justus von Liebig, like Laurent, noted that the "type" theory bore striking similarities to Laurent's already-maligned "nucleus" theory. Laurent himself charged Dumas with intellectual theft: "If it fails, I shall be the author, if it succeeds another will have proposed it. M. Dumas has done much for science; his part is sufficiently great that one should not snatch from me the fruit of my labors and present the offering to him."[3]

In fact there are significant distinctions between the two theories, particularly in their conceptual origins. For Laurent, the nucleus theory derived from his knowledge of mineral crystallography and chemical isomorphism in the tradition of René-Just Haüy (1743–1822) and Eilhardt Mitscherlich. Laurent's approach followed, then, from the natural history of classification of crystals. In contrast, Dumas's type theory was based in a natural history tradition that he had learned as a student in Geneva under the botanist Augustin-Pyrame de Candolle (1778–1841), who was the author of a botanical classification based on the concept of "type." Dumas's approach followed from the natural history of the classification of organisms.

Indeed, in the early 1830s, the natural history of organisms was at the center of a well-publicized dispute at the Paris Academy of Sciences between paleontologist Georges Cuvier (1769–1832), perhaps the most powerful figure of French science at the time, and zoologist Étienne Geoffroy Saint-Hilaire (1772–1844). The issue in question was what should be the basis for a "natural" system of classification in zoology.

One of Cuvier's fundamental premises in paleontological reconstruction of vertebrates had long been that an animal's needs and functions determine its structure. In contrast, pursuing a set of doctrines that he called "philosophical" or "transcendental" anatomy, Geoffroy argued that a generalized plan for vertebrate and invertebrate anatomy must be arrived at by studying the connections between parts and by determining homologies that relate similar parts in different organisms. Geoffroy's view implied that there is a fixed relation in the total number and arrangement of parts that is independent of the specific form of the parts and the uses to which they are put.

That Dumas was interested in this argument is not surprising, since it was the gossip of the Paris cultural elite. Dumas was well placed to make judgments because of his training under Candolle. Further, one of Dumas's closest friends in Paris was Henry Milne Edwards (1800–1885), the very man who began to develop a new type theory in order to work out a systematic compromise between Cuvier's and Geoffroy's extreme emphases on function and structure, respectively. Dumas clearly borrowed the type theory for chemical compounds from the natural history of living organisms, as well as from Laurent's mineralogical nucleus theory.

In a paper prepared for the Academy of Sciences and published in 1840, Dumas laid out a systematic language and theory for chemistry based in conceptions of the "type," the "genus," and the "natural family."[4] Liebig and Friedrich Wöhler were among those who opposed and ridiculed Dumas's (and Laurent's) new approach, with Wöhler writing a private satire that sounds almost like Jonathan Swift lampooning the seventeenth-century "projectors" of the Royal Society of London. Wöhler's piece was published by Liebig under the guise of a letter, allegedly from Paris, signed by S. C. H. Windler (the German for "swindler" is *Schwindler*) and appended by an unsigned footnote.

The alleged letter reported that all atoms but chlorine were eliminated following the substitution of chlorine in manganous acetate, but with the conservation of the acetate "type." The footnote added that fabrics spun entirely of chlorine were already for sale in London, "very much in demand in the hospitals and preferred over all others for night caps, drawers, etc."[4]

But despite his initial misgivings, Liebig was unenthusiastically conceding some value to the substitution theory by the 1840s, and, twenty years later, at a banquet in Paris over which Dumas presided, Liebig delivered a decidedly double-edged compliment. He had long since given up organic chemistry, he said, since, with the theory of substitution as a foundation, the theoretical work had been achieved and organic chemistry now only needed drudges for its completion.

From roughly 1840 to 1860, the substitution of elements or radicals into the molecular type became the accepted explanation for the production of compounds that could be understood as the primitive type's "derivatives" (or "degenerates," another term from natural history). Taking off from research of Dumas and Laurent, chemists extended type theory into a broader structure theory, including the concept that carbon is tetra-atomic or tetra-valued. This idea is the root of what came to be called valence theory. Liebig's influence can be seen in the fact that leading structuralists among French, German, and English chemists all studied, even if only briefly, with Liebig.

Of these, Charles-Fréderic Gerhardt and Charles-Adolphe Wurtz, both from Strasbourg, one Jewish and the other Protestant, made their careers in France. A. W. Hofmann, born in Giessen, moved to London in 1845 to direct the Royal College of Chemistry and then returned to Germany in 1865 to head the University of Berlin's laboratory of organic chemistry. Alexander Williamson, born in London of Scottish parents, studied chemistry under Thomas Graham (1805–1869) in London, then under Leopold Gmelin and Liebig in Germany, briefly attending Auguste Comte's course in mathematics at the École Polytechnique in Paris.

Wurtz, a protégé of Dumas, made methylamine and ethylamine in 1849 by treating the cyanic esters with potassium hydroxide. Using

Dumas's theoretical approach, he described his new compounds as molecules obtained by replacement of an equivalent of hydrogen with one of methyl or ethyl radical. The frequently debated hypothesis of the independent existence of the "radical" was given experimental confirmation by the experiments of Edward Frankland in London. He not only (erroneously) claimed to isolate ethyl but produced the new organometallic materials zinc ethyl and zinc methyl.

In 1847, Frankland and his German friend Hermann Kolbe together carried out various carboxylation and decarboxylation reactions that may have been the first reactions planned expressly to alter the carbon content of an organic molecule and to study molecular constitution. Kolbe's earlier synthesis in Germany of acetic acid was the first deliberately planned, step-by-step synthesis of an organic substance from purely nonorganic origins, a feat that Wöhler had not achieved in his accidental synthesis of urea in 1828.[5]

Shortly after he arrived in London (where he came to know both Frankland and Kolbe) in 1849, Hofmann's synthesis of new derivatives of aniline convinced him to abandon the dualistic theoretical approach (the "copula" theory) that was favored by Liebig and Kolbe. This dualistic theory attempted to interpret reactions as the coupling, or conjugation, of two entities, one passive and the other active. Hofmann's defection from this revisionist electrochemistry angered Kolbe, but Liebig managed to take pleasure in his former student's work, writing shortly after Hofmann arrived in London; "The new compounds of aniline are very interesting, namely the bases corresponding to ethylamide or amidethyl. You will certainly not regret having spent so much time working with aniline, because *the history and constitution* of such noteworthy compounds is *so valuable for theoretical chemistry*" [emphasis added].[6] Applying Hofmann's reasoning to the preparation of substituted alcohols, Williamson set out in 1850 to do experiments with ethyl iodide and potassium ethylate. The unexpected outcome was the production of diethyl ether and two important theoretical conclusions.

The first conclusion was Williamson's support for Gerhardt's recently published view that most textbook organic formulas should be halved, as would be required if standard atomic weights were taken to be C=12, O=16 rather than C=6, O=8. Thus, alcohol should be represented by C_2H_5OH, not $C_4H_{10}O.H_2O$. The Gerhardt system was largely adopted at the Karlsruhe congress in 1860, as discussed in chapter 2.

Second, Williamson proposed a new and simple "type." "Alcohol is . . . water in which half the hydrogen is replaced by carburetted hydrogen (C_2H_5), and ether is water in which both atoms of hydrogen are replaced by carburetted hydrogen"[7] Thus, ether and alcohol are both substituted water types.

Gerhardt integrated these results into his *Traité de chimie organique,* published in four volumes between 1853 and 1856. He died in 1856, before he turned forty. In his systematic textbook, Gerhardt added a hydrogen (H_2) type and a hydrogen-chloride (HCl) type to Williamson's water type and to Wurtz's and Hofmann's ammonia type. A fifth type, the methane (marsh gas; CH_4) type, was proposed in 1855 by William Odling (1829–1921) and in 1857 by Auguste Kekulé. (See figure 3.)

What then was really meant by the type theory? The "type" was a unitary structure in which an atom or group of atoms ("radical" or "residue") holds together other atoms or groups. Dumas distinguished between a "chemical type," which is not essentially altered

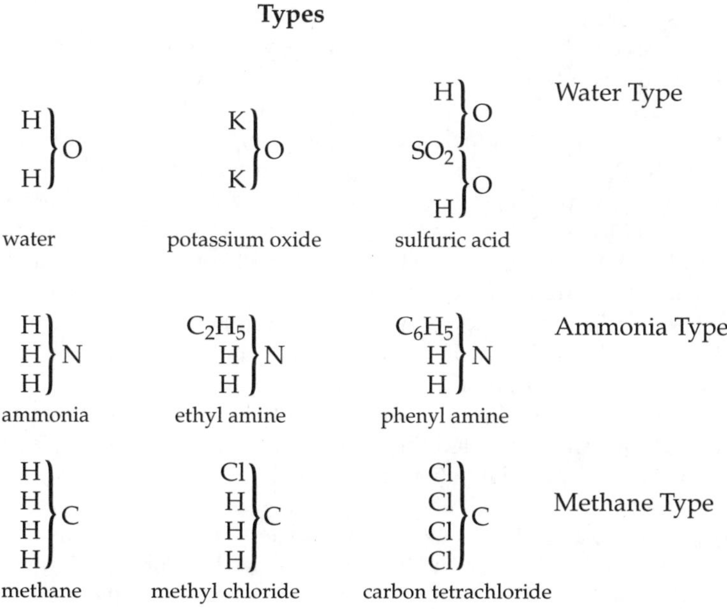

Figure 3. Three of the five molecular types proposed in the mid-nineteenth century

by substitution, and a "mechanical type," which is created by the addition or elimination of atoms. Arrays of hydrocarbons now could be predicted and explained as derivatives of a fundamental chemical type. For example, ethers, acids, acid anhydrides, aldehydes, and ketones were all derived from water by replacement of one or both hydrogen atoms by radicals or residues with explicit chemical functions.

For most chemists of Hofmann's and Frankland's generation, analogies between molecules and natural organisms seemed more appropriate than analogies between molecules and billiard balls or falling bodies. Many chemists were forsaking the mechanical language of earlier "chemical philosophy" and stressing the differences between physics and chemistry rather than their similiarity. Electrical explanation, in the tradition of Berzelius's electrodualism, all but disappeared from organic chemistry after 1860 and did not reappear until the turn of the century.

Constitution and the Structure Theory in Chemistry

As was common among chemists in the nineteeth century, August Kekulé traveled and studied widely in his youth. Born in Darmstadt in Hesse, he went to the University of Giessen in 1848 to study architecture. There he fell under the spell of Liebig and found himself under the supervision of Liebig's colleague Heinrich Will. In Paris from 1851 to 1852, Kekulé listened to Dumas lecture but spent some of his time talking with Wurtz and Gerhardt before returning to Giessen for his doctoral defense. For the next fifteen months Kekulé worked in Adolf von Planta's private laboratory in Switzerland. Then, thanks to an assistantship arranged by Liebig with his former student John Stenhouse at Saint Bartholomew's Hospital in London, Kekulé lived in London from 1853 to 1855. There he visited and argued with Hofmann, Frankland, Graham, and Odling, among others.

At an 1890 celebration in his honor, Kekulé reminisced about his youthful career and recalled the occasion atop a London bus (probably the summer of 1855), whence

> I fell into a reverie, and lo, the atoms were gamboling before my eyes! . . . A larger one embraced the two smaller ones; . . . still larger ones kept hold of three or even four of the smaller; whilst the whole kept whirling in a giddy dance. I saw how the larger ones formed a chain, dragging the smaller ones after them, but only at the ends of the chain.

> ... The cry of the conductor: "Clapham Road," awakened me from my dreaming; but I spent a part of the night in putting on paper at least sketches of these dream forms. This was the origin of the "Structural Theory."[8]

The theory, if dreamed in 1855, was not published until two years later. In the meantime, Kekulé passed his *Habilitationshrift* in Heidelberg and began teaching students in a ten-seat lecture room in a rented house outfitted with a small laboratory. Among his first students was Adolf von Baeyer, and among his later Heidelberg protégés were Emil Erlenmeyer, Friedrich Beilstein, Ludwig Carius, Lothar Meyer, and Henry Roscoe. Kekulé left Heidelberg in 1858 for the Flemish University of Ghent (where he lectured in fluent French) and then moved back to Germany, to Bonn, in 1867. At Bonn a mag-

Auguste Kekulé (*front row, center*) became head of one of the most important research schools in German chemistry when he became director of the Chemical Institute at Bonn in 1865. *Courtesy of the Edgar Fahs Smith Collection at the University of Pennsylvania.*

nificent chemical laboratory had been negotiated by Hofmann, who was ready to leave London but ended up taking a professorship at Berlin. Kekulé, chosen by default, was enormously successful at Bonn. Indeed, three of the first five Nobel Prizes in Chemistry were awarded to his Bonn students, including van't Hoff (in 1901), Emil Fischer (1902), and Baeyer (1905).

For Kekulé's research "school"—a term increasingly commonly applied to laboratory research groups in the sciences during the nineteenth and twentieth centuries—as for almost all research schools in late nineteenth-century organic chemistry, the leading theoretical paradigm was the structure and valence theory, along with the benzene-ring chemistry of aromatic hydrocarbons, all closely associated with Kekulé's name.

It was during the years 1857–1858 that Kekulé developed the theory that carbon atoms link together by using up two of their four affinity (or atomicity) values, just as hydrogen, oxygen, and nitrogen atoms can link together through their affinity units or values. The carbon "chain" was described as a "backbone" or "vertebral column" (Kekulé used the word "skeleton" in 1858) through which homologous substances can be built up. As the column lengthens in the simplest possible way, for example from CH_3Cl to C_2H_5Cl to C_3H_7Cl, the properties of the substance change in a systematic way. A gas, condensing at 23 degrees, becomes a liquid whose boiling point is 26.5 degrees, and then a liquid whose boiling point is 46.5 degrees. Of all the elements, carbon is the most prolific of atoms in giving birth to arrays of new compounds. As Kekulé put it, "Carbon is tetratomic, . . . [and] it enters into combination with itself."[9]

The Scotsman Archibald Scott Couper (1831–1892), who was working in Wurtz's Paris laboratory at the time, published the same ideas a few weeks after Kekulé's influential 1858 paper. Couper later blamed Wurtz for delaying the publication of his paper. Unlike Kekulé, who used rounded or "sausage" graphical formulas for carbon linking, as shown in figure 4, Couper first designated the linking

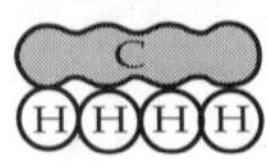

Figure 4. August Kekulé's "sausage" or rounded formula for representing the linking of atoms in the methane molecule

by dots, then switched to straight lines. These lines quickly gave rise to the somewhat ambiguous language of chemical "bonds."

The term *chemical structure* was popularized by the Russian chemist Aleksandr Butlerov (1828–1886), who met Kekulé in Heidelberg as well as Couper and Wurtz in Paris in 1857 while he was on leave from the University at Kazan in eastern Russia. Butlerov used the term *structure* to mean arrangement of atoms within the molecule—in other words, the attachments of the molecule's components to the carbon backbone. By 1861 Butlerov had the firm idea that this arrangement, or "constitution," was the cause of the molecule's physical and chemical properties and that structure accounted for what Kekulé and many chemists still called the "metamorphoses" of types.[10]

By 1864 Alexander Crum Brown (1838–1922), a former student of Kolbe and of Robert Bunsen and a professor of chemistry in Edinburgh, was using a graphical notation, similar to Couper's, for the constitutional or structural formula that was so convenient that it was adopted and has been used almost unaltered to the present day (see figure 5).

The power of these structural representations was immediately apparent. They made possible a kind of predictive guesswork that Kolbe disparagingly called "paper and pencil chemistry." It allowed a chemist to predict both derivatives and isomers, based on the equivalent atomicities or valences of atoms, without leaving the desk. The graphical possibilities were endless and they provided innumerable hypotheses for laboratory investigations. The challenges now lay in realizing the synthesis of possible compounds and in trying to explain why some compounds simply do not exist. "Vinyl alcohol" (CH_2=CH-OH), for example, could not be produced as a stable compound no matter how skilled the chemist.

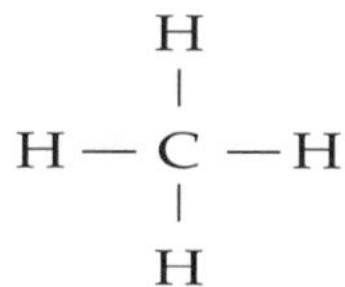

Figure 5. Constitutional or structural formula using straight-line notation for representing linking of atoms in the methane molecule

Kekulé's most famous publication was his paper of 1865 "On the Constitution of Aromatic Substances." Here he proposed the theory that all aromatic compounds, including the simplest one (benzene, then called benzol), contain a six-carbon "nucleus." This nucleus was represented by a closed chain of alternating double and single bonds. Laurent, the inventor of the "nucleus theory," had proposed in 1854 a hexagon crystalline form for benzoyl chloride.

Between 1866 and 1868 the simple "hexagon" figure became a standard representation for Kekulé's cyclohexatriene structure. It

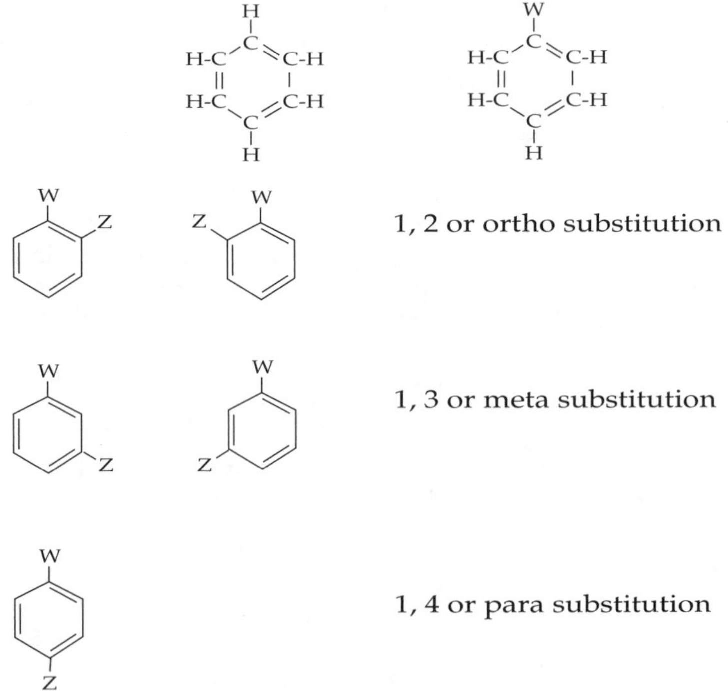

Figure 6. *Ortho, meta, para* patterns of substitution in aromatic hydrocarbons like benzene. *W* represents the position of the first substitution of an element for hydrogen and *Z* the position of the second substitution, which is determined by the identity of *W*.

was invented by Kolbe's former student Adolf Claus (1840–1900) and speedily adopted by Hofmann, among others. One of Kekulé's students, Wilhelm (Guglielmo) Körner (1839–1925), demonstrated decisively by 1874 the equivalence of benzene's six hydrogen atoms. He also established a clear laboratory routine for deciding whether di-substituted isomers were *ortho, meta,* or *para* (see figure 6). The synthesis and elucidation of the aromatic compounds and their derivatives remained a major theme of organic chemistry for the next fifty-plus years.

Stereochemistry and Molecular Architecture

Most schoolchildren of the world, not just French schoolchildren, know of Louis Pasteur (1822–1895). Readers of late-twentieth-century scientific and medical journalism, especially those following progress on development of an AIDS (in France, SIDA) vaccine, know of the Pasteur Institute, a research institution founded in 1888 in a country where the prestige of Pasteur's name rivals that of the general and emperor Napoleon.

Arguing an analogy between fermentation and disease, Pasteur proved that living microrganisms are responsible for both good wine and deadly cholera. Enlisted to advise the beer, wine, and vinegar industries of the regions where he lived, he developed the procedure for heating that came to be called "pasteurization" of beverages. His study of microorganisms resulted in the saving of the valuable silkworm industry in southern France, as well as in the further development of the vaccine technique discovered in 1796 by English physician Edward Jenner for preventing small pox. Perhaps most famously, Pasteur isolated and attenuated the strength of specific viruses or bacteria for the treatment and prevention of sheep anthrax (late 1870s), chicken cholera (early 1880s), and rabies (mid-1880s).

Pasteur also made considerable contributions to chemistry. Educated in Arbois and Besançon, he became a student of Dumas and A.-J. Balard (1802–1876), as well as a laboratory associate of Laurent in the early 1840s. Pasteur defended his doctoral thesis in 1847 and from 1848 to 1857 taught in Dijon, Strasbourg, and Lille (where he was dean of the sciences faculty). His modus operandi in all these locales—the development of close professional and personal ties with the regional commercial bourgeoisie—was motivated both by his own personal interests and by the French ministry of education's strong encouragement to find useful applications for scientific research. During 1857 to 1867 Pasteur was director of

PASTEUR EN 1852, A STRASBOURG

Louis Pasteur at the University of Strasbourg in 1852. Hanging from a ribbon around Pasteur's neck is a magnifying glass for separating crystals of the isomeric salts of tartaric acid from the optically inactive racemate. *Courtesy of the Edgar Fahs Smith Collection at the University of Pennsylvania.*

scientific studies at the École Normale Supérieure in Paris, one of the elite *grandes écoles* that then admitted only ten or fifteen science students a year. Under Pasteur's leadership the school came to rival the more distinguished École Polytechnique in the recruitment and training of mathematicians and scientists.

Largely as a result of his close association with Laurent from the period they both worked in Balard's laboratory, Pasteur chose a study of crystals, including the crystal tartrates, as the focus of one of the two subjects required for his doctoral thesis. His choice was also influenced by the interests of Jean-Baptiste Biot, whose studies of optically active substances included solutions of the salt of tartaric acid.

Biot, then professor of mathematical physics at the Collège de France, had just received in 1840 the Rumford Medal of the Royal Society of London for his experiments on the optical activity of solutions.

Biot and his colleague François Arago found that a monochromatic beam of plane-polarized light is twisted or rotated by solutions of sugar, camphor, or tartaric acid, implying that an asymmetrical structure lay not in the arrangement of molecules with respect to one another, but in the very interior of a single molecule. To study this phenomenon, an instrument called a "polarimeter" was devised, using a fixed and a movable Nicol prism at either end of a tube and enabling the experimenter to make readings distinguishing between *dextro* (right) rotatory solutions and *levo* (left) rotatory solutions.

Tartaric acid is the constituent of grapes that produces the sludge commonly called tartar (potassium acid tartrate) that accumulates at the bottom of wine casks. Pasteur studied a group of tartrates, investigating their chemical formulas and their similarity (isomorphism) or dissimilarity in crystalline form. He became interested in the hemihedrism, or asymmetry, of the crystal faces, as had Laurent and Eilhard Mitscherlich before him. Observing what he later termed right-handedness in the crystal faces of sodium-ammonium tartrate, Pasteur was surprised to find that some crystals of sodium-ammonium paratartrate were oriented to the right, others to the left, and some, he intitally thought, gave a symmetrical pattern.

In a study of Pasteur based on Pasteur's laboratory notebooks, historians Gerald Geison and James Secord have traced how Pasteur came to recognize the importance of an anomaly between what he was observing and what he was expecting. He knew that Mitscherlich had reported in 1844 that sodium-ammonium paratartrate (racemic acid salt) and sodium-ammonium tartrate are isomorphic and of the same chemical formula, but that the first compound does not deviate the plane of polarized light and the other does so. Yet Mitscherlich claimed that the paratartrate (racemate) crystallizes into a salt indistinguishable from the salt of tartaric acid.

Pasteur sought to improve his method of crystallization in order to obtain larger crystals. The temperature of his racemate solution was cool enough (below 27° C, a transition point) that his crystals were more clearly hemihedral in appearance, some with faces oriented in one direction and others with faces in the opposite direction. Had crystallization occurred above this temperature, the results would not have been so clearcut.

Separating the differently oriented crystals by hand, Pasteur found that their solutions were both optically active and that one set of crystals, matching the visual asymmetry of the tartrate crystals, was *dextro*-rotatory, while the other was *levo*-rotatory. Physically and visually, the two sets of crystals were mirror images of one another. Before he would present Pasteur's results at the Academy of Sciences, Biot insisted that Pasteur repeat the crystallizations with reagents that Biot provided, in Biot's own laboratory, using a sample of racemic acid that Biot had previously found optically inactive. All turned out as Pasteur predicted.

Pasteur continued his research on the tartrates for about six years before he began to commit more of his time to fermentation research. His experiments included testing whether electromagnetic forces influence asymmetries in crystallization. During the years 1858 to 1860, he discovered that the mold *Penicillium glaucum* consumes the *dextro*-rotatory molecules of racemic acid much more readily than the *levo*-rotatory molecules. In reporting these results he speculated on the possible asymmetric structures of the molecules in three dimensions. "Are the atoms of the *dextro* acid grouped on the spirals of a right-handed helix or situated at the corners of an irregular tetrahedron, or have they some other asymmetric grouping?"[11] He speculated that the forces of life may be inherently asymmetrical and he resisted the argument that yeasts and ferments act by virtue of their specific chemical natures rather than by virtue of a living force associated with asymmetry. In these views, Pasteur's work pointed back to the vitalists rather than allying him with his French contemporaries Berthelot and Wurtz.

Pasteur's idea that atoms in an organic molecule might be grouped in helices was to return in Linus Pauling's alpha-helix protein structure, as well as in the double-helix structure proposed by Francis Crick and James Watson for deoxyribonucleic acid (DNA) in the early 1950s. The speculation that atoms of a hydrocarbon might be situated at the corners of a tetrahedron more immediately reappeared briefly in Butlerov's speculation in 1862 that the magnitude of affinity forces might vary in the different directions defined by the apexes of a tetrahedron. Apparently based on his reading of Butlerov's paper, Kekulé proposed in 1867 that a three-dimensional representation using tetrahedra would better explain triple bonds than does a two-dimensional representation. Double and triple bonds had appeared in papers by Loschmidt (1861), Erlenmeyer (1862), and Crum Brown (1864).

Kekulé's speculation was well known to his students, among them the Scottish chemist and physicist James Dewar (1842–1923), who in 1867 began designing wire models bent into shapes representing compounds that might result from tetrahedral carbons. At Kekulé's laboratory in Bonn during the winter of 1872 to 1873, the Dutch student van't Hoff picked up on these ideas; later, in the spring of 1874 when he was in residence in Wurtz's Paris laboratory at the same time as Joseph-Achille Le Bel (1847–1930), van't Hoff became acquainted with Pasteur's views on the subject.

As was mentioned in chapter 4, in 1874 van't Hoff and Le Bel published papers (in Dutch and French journals, respectively) proposing that when a carbon atom is attached to four different atoms or groups, the four substituents can be arranged in two different ways as mirror images of one another. Van't Hoff explicitly proposed the geometry of the tetrahedron. He also drew on Johannes Wislicenus's recent publication on the two isomers of lactic acid ($CH_3CH(OH)CO_2H$) which Wislicenus had shown to be structurally identical despite their dissimilar properties. The only possible explanation, Wislicenus wrote, was that "the difference is due to a different arrangement of the atoms in space."[12]

Once he saw van't Hoff's French pamphlet, Wislicenus asked his student Felix Herrmann to prepare a German translation, for which Wislicenus wrote the preface and arranged publication with the German science publisher Vieweg Verlag. News of this publication precipitated the infamous attack by Kolbe (mentioned in chapter 2), who in any case was in bad humor, having just lost out to Kekulé's student Baeyer in competition for Liebig's succession at Munich. Thus, Kolbe directed his ire at another Kekulé protégé, van't Hoff, heaping scorn on van't Hoff's "cosmic space."[13] Wislicenus was among many German chemists who thought Kolbe had gone too far. Kolbe's attack backfired and tarnished his own historical reputation, particularly after van't Hoff reprinted Kolbe's diatribe in the second edition (1887) of his *Chemistry in Space.*

In addition to the hypothesis of the carbon tetrahedron, in his 1874 paper van't Hoff also dealt with the kind of isomerism that had been noted for fumaric acid and maleic acid, in which there are two, not four, dissimilar substituents. The difference in properties of these identical constitutional molecules, he suggested, might be due to the impossibility of free rotation around a double bond.

In the late 1880s Wislicenus demonstrated that van't Hoff's interpretation was correct and that the carboxyl constituents are on opposite sides of the double bond in fumaric acid and on the same side in maleic acid (see figure 7). In 1892, Baeyer introduced the terminology *cis/trans* for the geometrical isomers on the grounds that "maleoid," "fumaroid," and other labels were awkward (figure 7). Earlier, in 1885, Baeyer had proposed a theory of "strain" for planar carbon rings, based in notions of stereochemistry. Five- and six-membered rings are more stable than other ring sizes, he hypothesized, because distortions from the tetrahedral angle of 109^0 are small. While van't Hoff was largely to abandon organic chemistry in favor of physical chemistry after 1880, Baeyer, along with his student Emil Fischer, was to become one of the preeminent organic chemists of Germany by the early 1900s. They all, as mentioned earlier, were associates of Kekulé.

Biological Molecules and Physical Methods

Adolf von Baeyer had made his reputation by the early 1880s for his work in synthetic dyestuffs based squarely in his knowledge of structural chemistry. A great achievement of his research group was the synthesis of alizarin, the next dyestuff synthesized after William Henry Perkin's accidental production of mauve when he was eighteen. Mauve became the first commercial aniline dye. Perkin himself

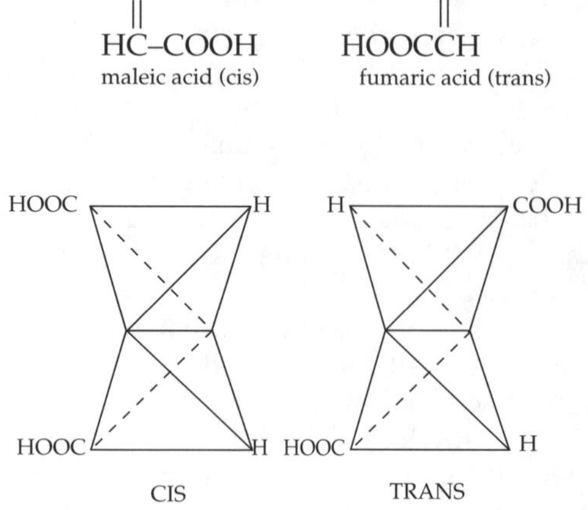

Figure 7. Maleic and fumaric acid: *Cis-trans* geometrical isomers

became the first synthetic dye manufacturer, much to the initial regret of his teacher Hofmann, who wanted Perkin to aspire to an academic career in chemical research.

Baeyer's success with alizarin was followed in 1880 in Munich by his synthesis of indigo, a natural product of the woad plant farmed in India. Baeyer sold both processes to industry, where they were so improved and successfully marketed that by 1900 the natural indigo dye had begun to disappear. Baeyer was typical of German chemists who maintained close connections with industry, especially BASF (Badische Anilin und Soda-Fabrik), where the chemist Heinrich Caro (1834–1910) worked after 1868 and became head of the first formally designated industrial research laboratory in 1877. As an effective teacher whose research included significant contributions to chemical theory, Baeyer exemplifies the successful amalgamation of "pure" and "applied" science.

The same can be said of his most successful pupil Emil Fischer (1852–1919). Fischer moved in his research interests from the production and characterization of relatively simple aromatic compounds in the 1870s and 1880s to the study of increasingly complex molecules of biological significance, including the sugars, purines, and proteins. During the 1880s he was offered Caro's position at BASF at an extraordinary salary of 100,000 marks (approximately three times Kolbe's income at Leipzig in 1875). But Fisher turned down the BASF offer.

Instead he taught as full professor (*ordentlicher Professor*) first at Erlangen (from 1882) and then at the University of Würzburg (from 1885), moving in 1892 to the pinnacle of German academic chemistry by succeeding Hofmann at the University of Berlin. Around 1905 at Berlin, Fischer's research group included some twenty-five to thirty assistants, advanced students, and guests. By the end of his career, his published collected works included 600 experimental articles, of which 120 were published under the sole authorship of his Ph.D. students, but clearly showing Fischer's direction.

Fischer's work on sugars remains fundamental to the study of carbohydrates: those natural substances, including sugars, with the empirical formula $C_nH_{2n}O_n$ or $C_n(H_2O)_n$. Fischer began his investigations of the sugars in 1884 when it was known that all sugars contain either the aldehyde or ketone functional group. Upon hydrolysis, for example, cane sugar (sucrose) yields two C_6 sugars (or hexoses): glucose and fructose, glucose being an aldehydic sugar (aldohexose) and fructose a ketonic sugar (ketohexose).

Reasoning from stereochemical considerations, Fischer concluded that glucose must have four asymmetrical carbon atoms and that, based on van't Hoff's hypothesis, the number of possible isomers must be 2^4 or 16, consisting of 8 pairs of enantiomorphs. Fructose, with three asymmetrical carbon atoms, would have 2^3, or 8, isomers.

In addition to glucose, only two other aldohexoses, mannose and galactose, were known to Fischer. Fructose is the only common natural ketohexose. Using phenylhydrazine ($C_6H_5NHNH_2$), a compound he had discovered in 1875, to form the crystalline osazone with the sugars, Fischer found that glucose, fructose, and mannose all produced the same osazone. This result gave him a clue that these three sugars differed in arrangement of substituents on their first two carbon atoms, but not the "bottom" four, as shown in figure 8. Fischer successfully identified the isomeric structures of fructose, mannose, and glucose and continued his investigations by characterizing and synthesizing the remaining fourteen aldohexoses and seven ketohexoses.

This he accomplished from 1891 to 1896, using not only chemical techniques based on the structure theory but also techniques from analytical and physical chemistry including, of course, optical activity. Procedures worked out by Heinrich Kiliani (1855–1945) were valuable in the work on sugars, especially an addition reaction of hydrogen cynanide to sugars, which Fischer used for lengthening carbon chains. Fischer studied the splitting of glycosides by enzymes, leading him to the now well-known "lock and key" analogy for the specificity of enzyme action.

Fischer's study of caffeine in 1881 resulted in the demonstration that caffeine, uric acid, xanthine, and guanine are all derivatives of a substance he named purine. Baeyer had begun to map out some of

Glucose	Fructose	Mannose
CHO	CH_2OH	CHO
H–C–OH	CO	HO–C–H
HO–C–H	HO–C–H	HO–C–H
H–C–OH	H–C–OH	H–C–OH
H–C–OH	H–C–OH	H–C–OH
CH_2OH	CH_2OH	CH_2OH

Figure 8. The structure of the three sugars glucose, fructose, and mannose as determined by Emil Fischer

these relationships in the 1860s. The work on the nitrogen-containing purines dovetailed with increasing attention to proteins, which contain nitrogen in addition to carbon, hydrogen, oxygen, and usually sulfur. The study of proteins was without doubt one of the most difficult and complex areas of organic chemistry in the late nineteenth century. As their name suggests, proteins (from Grk., *protos,* meaning "first") were considered the key to life.

Proteins were even more notorious than most organic compounds for their fragility when heated. Some, like fibroin, collagen, and keratin are insoluble in water; others, including egg albumin, casein, and plasma proteins, are soluble in water or in aqueous solutions of acids, bases, or salts. In all cases, gentle methods of analysis are best, including the use of mild alkalis and acids.

In 1820 Henri Braconnot had used the hydrolysis of gelatin to make a sweet-tasting substance, which he named "sugar of gelatin." In 1846 E. N. Horsford, working in Liebig's laboratory, correctly identified the sugar's nitrogen content, and Berzelius (who died in 1848) renamed the substance glycine. By the late nineteenth century, when Fischer and his laboratory school were turning to protein chemistry, he had worked out the structure of cellulose as a polysaccharide, meaning that it was a chain of glucose units, and he was ready to apply a linking hypothesis to proteins as well.

Here again Kekulé's influence can be seen. Kekulé's inaugural address as rector of the University of Bonn in 1877 included remarks about the natural organic substances most closely responsible for biological processes. Proteins, starch, and cellulose, he suggested, may consist of very long chains and may take their special properties from unusual structures: "A considerably large number of single molecules may, through polyvalent atoms, combine to *net-like,* and if we like to say so, *sponge-like masses,* in order thus to produce those *molecular masses* which resist diffusion, and which . . . are called *colloidal* ones."[14]

What form would the linkages take? At the 1902 annual meeting of the Gesellschaft der deutscher Naturforscher und Ärtze (Society of German Scientists and Physicians), Fischer and Franz Hofmeister (1850–1922) of Strassburg independently proposed that proteins are constituted by amino acids joined together by the condensation of the amino group (NH_2) of one amino acid with the carboxyl group (COOH) of another, forming amide bonds (–CONH–). Using a method pioneered by Theodor Curtius (1857–1928) for the formation of the dimer 2,5 diketopiperazine through the elimination of alcohol from glycine ester, Fischer boiled the dimer with concentrated hydrochloric acid to form the glycylglycine through a –CONH– link:

$$H_2N–CH_2 \cdot CO \cdot NH \cdot CH_2COOH$$

Fischer and his group went on to combine more than two amino acids, moving from what Fischer called "dipeptides" to "polypeptides." In 1907 they experienced the triumph of constructing an eighteen-amino-acid polypeptide: leucyl-triglycyl-leucyl-triglycyl-leucyl- octaglycyl-lycine. These polypeptides behaved like intermediate products of protein hydrolysis.

How big, then, would a completely synthesized protein be? François-Marie Raoult's method of inferring molecular weight from the depression of the freezing point of a solution was used by Horace T. Brown and G.H. Morris in 1889 to estimate the molecular weight of natural "soluble starch " at 32,400. Fischer explicitly rejected a reported estimate of molecular weight for natural proteins (including hemoglobin) of 12,000 to 15,000 although he accepted as valid the figure of 4,021 for a starch derivative that he synthesized with Karl Freudenburg (1886–1983). Fischer did think, however, that natural proteins were likely to have smaller weights than artificial ones.

About this time, in 1915, Wilhelm Ostwald's son Wolfgang Ostwald (1883–1943) published a popular work with the catchy title *The World of Neglected Dimensions* (*Die Welt der vernachlässigten Dimensionen*). The younger Ostwald supported and even lowered Fischer's cap on molecular weight, laying out principles of what came to be called the aggregate theory of colloids. Along with Ostwald, Carl Harries and Rudolf Pummerer (studying rubber), Kurt Hess and Paul Karrer (cellulose), and Max Bergmann (proteins) subscribed in the 1920s to a physico-chemical theory that colloidal substances such as cellulose, rubber, starch, proteins, and resins are the physical aggregates of small molecules held together by some intermolecular force, perhaps akin to Alfred Werner's "secondary valence" or Johannes Thiele's "partial valence."

In 1920, a year after Fischer's death, Freudenberg offered new experimental data from cellulose degradation that cellulose is a long-chain structure with very high molecular weight. The next year Michael Polanyi, a Hungarian physical chemist working in the laboratory directed by Reginald Herzog at the Kaiser Wilhelm Institute for Fiber Chemistry, presented a lecture commenting on recent X-ray analysis by Herzog and William Jancke of cellulose samples.

In his lecture Polanyi concluded that the measured X-ray diffraction spots were consistent with either of two structures. The cellulose was either a long glucosidic chain or an aggregate made up of units of a small number of glucose anhydrides. Polanyi declined, however, to rule one way or the other on the basis of the physical evidence alone. Chemists at the Berlin colloquium who heard this lecture strongly favored lower molecular weights for natural biological molecules and regarded Polanyi as an interloper in matters having to do with organic chemistry.

Polanyi did not know at this time about Freudenberg's new estimate for the higher weight of cellulose, nor was he familiar with the recent argument for large molecules advanced by Hermann Staudinger (1881–1965) of the Zurich Polytechnic Institute (Eidgenössische Technische Hochschule). In 1922 Staudinger provided strong chemical evidence for very long chains by refuting predictions from Harries's aggregate theory of rubber. According to Harries, partial valences generated from unsaturated double bonds hold together relatively small molecules in rubber; thus hydrogenation of rubber should yield a low molecular substance. Staudinger and his associate Jakob Fritschi found, in contrast, that properties of hydrorubber were not those of a distinct, low-weight molecular substance, but were similar to those of natural rubber. They used the term *Makromoleküle* for the first time.

Debates about the structure and weight of natural substances such as cellulose and protein demonstrate the rivalry and misunderstanding that was hardly unusual among organic chemists, physical chemists, and physicists in the late nineteenth and early twentieth centuries. The discovery of X rays in 1895, discussed in the next chapter, gave rise within twenty years to the kind of X-ray diffraction studies of molecular structure discussed by Polanyi in his failed Berlin colloquium lecture.

Just as some chemists in the late nineteenth century resisted the identification of new elements by spectroscopy or the estimation of molecular weight from freezing-point depression, many chemists initially were unenthusiastic about X-ray data. Many thought their suspicion justified in 1926, when X-ray data on atomic dimensions was used by L. O. Sponsler and W. H. Dore to argue an alternating 1–1, 4–4 bonding betweeen glucose units in cellulose even though this interpretation contradicted irrefutable chemical evidence of 1–4 bonding.

Undeterred, Hermann Mark (1895–1992) was among those who pioneered the application of X rays in natural-product chemistry.

Working with Kurt Meyer (1883–1952), Mark developed a theory of long-chain molecules held together in "banks" or "micelles," using the widths of X-ray diffraction spots to estimate the size of the micelles at 50 primary chains wide with each chain composed of some 50 glucose units. By the mid-1930s estimates of macromolecular weights as high as a million were supported by both physical and chemical investigations. Theodor Svedberg provided weight estimates in his Uppsala laboratory with the ultracentrifuge, while the Du Pont industrial research chemist Wallace H. Carothers (1896–1937) confirmed the existence of macromolecules and their constitutive mechanism of polymerization.

Among those who learned Mark's methods was Linus Carl Pauling (1901–1994), who visited Mark in Germany in 1930. Mark's use of X-ray diffraction and electron diffraction intrigued Pauling, as did Mark's theories of protein structure and the flexibility of polypeptide chains. In 1932 Pauling, already well known for his application of quantum wave mechanics to the chemical electron bond (see chapter 6), sought funding from the Rockefeller Foundation to support his research on structural chemistry. This work would include investigations of the structures of "protein, hemoglobin and other complicated organic substances."[15] When the foundation shifted its interest more squarely from chemistry and physics to biology and medicine, Pauling targeted hemoglobin for study in his grant application of 1934. Pauling was becoming a "biochemist."

Using techniques of crystallization of amino acids and X-ray investigations of their dimensions, Pauling began building up pictures of the way the protein must look. From 1937 until 1951 he and his coworkers at Caltech discovered the role of hydrogen bonding and assigned positions to amino acids on a coiled chain, using measurements of interatomic distances and bond angles, as well as three-dimensional molecular models, to develop a representation of what became known as the alpha helix. Not only the constitutional structure but the three-dimensional architecture of the molecule became a key to the chemistry of life.

The alpha helix was announced in 1950. Francis Crick and James Watson published the double-helix structure for DNA in 1953. However, as the chemist and historian of chemistry John Hudson notes, perhaps a more dramatic turning point in structure determination came earlier, in 1944, when Dorothy Crowfoot Hodgkin (1910–1994) and her research group at Oxford completed the determination of the structure of penicillin (molecular weight 334) by X-ray diffraction methods well before organic chemists had agreed on a structure.

Hodgkin, who received the Nobel Prize in Chemistry in 1964 for her structure determination of vitamin B_{12}, directed a research laboratory at Oxford where women made up as much as one-third of the students and staff. The later British prime minister Margaret Thatcher was one of her students. In 1947, Hodgkin was elected a Fellow of the Royal Society, where she joined five other women in membership. The X-ray crystallographer Kathleen Lonsdale (1903–

Dorothy Crowfoot Hodgkin won the Nobel Prize for Chemistry in 1964 for her determination of the structure of vitamin B_{12}. After studies at Oxford, she worked with J. D. Bernal at Cambridge and produced the first X-ray diffraction photograph of the protein pepsin in 1934 before taking her doctorate in 1937. *Courtesy of The Royal Society, London and, for reproduction permission, Godfrey Argent Ltd.*

1971) and the biochemist Marjorie Stephenson (1885–1948) were the first women elected to the Society, in 1945.

At the time that Hodgkin announced her structure for penicillin, probably a thousand other chemists in Great Britain and the United States were working on the problem of penicillin, some of them also at Oxford in a group headed by one of the acknowledged masters of the chemistry of natural products, Robert Robinson. Robinson did not initially accept Hodgkin's results, partly because of his disinclination toward and distrust of physical instrumentation, including the apparatus of X-ray crystallography.

Robinson, who received the Nobel Prize in Chemistry in 1947 for his investigations on alkaloids and other plant products, represented in the mid-twentieth century the highest standards of practice for organic chemistry in the "old school" of the natural history tradition. His disinclination for spectrometers does not mean he was close-minded to new physical ideas, however. As will be discussed in chapter 6, he was innovative and influential in thinking about the usefulness of ions and electrons for understanding the cause of chemical affinities and the mechanisms of chemical reactions.

By the 1960s many chemists of Robinson's generation were expressing some nostalgia for the good old days when chemistry was *chemistry,* not spectroscopy and calculating machines. But they all understood that the applications of the new physical instruments only confirmed the explanatory power and predictive accuracy of nineteenth-century structural chemistry in the natural history tradition. This chapter has stressed the relationship between chemistry and natural history; the next will emphasize the relationship between chemistry and physics and, in particular, the revival of an electrical theory of chemical affinity within a twentieth-century framework of electron dynamics.

6

A New Chemistry, a New Physics: Radiations, Particles, and Wave Mechanics after 1895

The "Naughty" Röntgen Rays

One of the single most dramatic discoveries in the history of science took place in a physical laboratory at the University of Würzburg in early November 1895. On that evening, Wilhelm Conrad Röntgen (1845–1923) was working with a Crookes-type vacuum discharge tube that he had been slowly evacuating for several days. The tube, covered by a lightproof black cardboard jacket, was set up not far from a sheet of paper coated with barium platino-cyanide. This sort of screen served experimentalists as a detector by fluorescing when it was struck by visible light or by the radiations that Philipp Lenard had recently discovered are emitted through an aluminum-foil window sealed into the discharge tube.

Röntgen's discharge tube had no window. The tube was covered. Yet to his surprise he noted that in the darkened room the screen was fluorescing as if it were being struck by radiations. Indeed, when he turned the screen around, the side without the coating also fluoresced. Objects placed between the tube and the screen were transparent to the invisible radiations. Then, astonishingly, he found that when he placed his hand between the tube and the screen, the darkened image of the bones of his hand appeared on the screen.

Working quietly for seven weeks in his laboratory, and telling his wife that he was on to something that was either important or foolish, Röntgen established to his satisfaction that the rays, like light, propagated in straight lines, but that unlike light they showed no refraction, reflection, or diffraction that he could detect. Nor were they deflected in a magnetic field, as cathode rays were. He found that in addition to causing fluorescence, the rays affected photographic plates in ways that were totally new. He used the rays to make photographs of a compass needle through its metal case, nails in the door frame of a closed room, a key on a pile of photographic plates, and the bones of his fingers.

Röntgen's confidence about the validity of his discovery grew, and at the end of December he wrote a paper for the Physico-Medical Society of Würzburg. On New Year's Day of 1896 he mailed reprints and samples of some of the *X-Strahlen* photographs to colleagues throughout the world. On 5 January 1897, a Vienna newspaper announced the discovery. The London *Daily Chronicle* and the New York *Sun* followed the next day. Scientists in laboratories everywhere immediately began duplicating his results, and on 13 January Röntgen nervously gave a command demonstration for Kaiser Wilhelm II at the court in Berlin. During the next year more than five hundred articles and books appeared about the new penetrating radiations. As one wit penned:

> I hear they'll gaze
> Through cloak and gown—and even stays,
> These naughty, naughty Röntgen rays.[1]

It did not take much imagination to recognize that *X-Strahlen* could be used to photograph bullets in flesh and fractures in bones. Their medical application was immediate in what was likely the quickest-ever diversion of pure science into technology.

What were these mysterious "X" rays? Röntgen, who continued investigations of them only through 1897, postulated that they were the longitudinal (compression) ether waves previously predicted by the elastic-solid model of the ether. G. G. Stokes and J. J. Thomson were among those who preferred an alternative "impulse" hypothesis, in which cathode rays were assumed to engender spherical ether waves on impact with glass or metal surfaces. These pulses in turn collided with material atoms as if the pulses were particles.

Both the compression and pulse theories lost ground after Dutch scientists Hermann Haga and Cornelis Wind demonstrated that there was good evidence for X-ray diffraction and, in England, Charles G. Barkla reported the polarization of X rays. These investigations occurred during the period from 1899 to 1904. In 1912, Walter Friedrich and Paul Knipping confirmed in Munich, where Röntgen had moved from Würzburg, that Max von Laue's theoretical prediction that diffraction phenomena would be produced by passing X rays through the natural three-dimensional diffraction grating of a crystal was correct. Thus X rays were transverse wave vibrations some 10^{-4} times smaller than visible light.

In the summer of 1912 William Henry Bragg (1862–1942), who had previously favored the pulse theory of X rays, discussed Laue's work with his son William Lawrence Bragg (1890–1971), who was studying at Trinity College, Cambridge. The younger Bragg began experimental work that led him to deduce the "Bragg equation," which relates angles of X-ray diffraction to X-ray wavelength and to the distance between atoms in a crystal. The two Braggs spent their vacations together using the X-ray spectrometer designed by the elder Bragg in his Leeds laboratory to determine atomic arrangements in crystals, including diamond. For this work they became the first and only father-and-son team to receive a Nobel Prize in Physics, in 1915.

In these same years, Henry H. G. J. (Harry) Moseley (1887–1915) and George Galton Darwin (1887–1962) began investigating X-ray diffraction and its relationship to atomic properties in Ernest Rutherford's laboratory at Manchester. Returning to Oxford, where he had earlier begun his university studies, Moseley completed a series of investigations that revealed a regular shift in X-ray wavelengths emitted by a series of chemical elements when each was struck with electrons in an X-ray tube. Each element was given a number, beginning with 1 for hydrogen, proportional to the square root of the frequency of one of the X-ray lines. A straight line could then be graphed by plotting the square root of the reciprocal of the wavelength versus this series of "atomic numbers."

This method provided predictions of new elements. Unknown atomic numbers corresponding to unknown elements could be obtained by interpolation. Moseley successfully identified as new elements some rare-earth samples brought to Oxford by the French chemist Georges Urbain (1872–1938), thereby confirming the usefulness of this new physical method for what had previously been a problem of chemical analysis. Soon atomic number replaced atomic weight as the organizing principle for the periodic table of the elements. Moseley correctly predicted that there must be ninety-two natural elements up to and including uranium, as well as fourteen rare-earth elements.

But what was the meaning of "atomic number," a number smaller than atomic weight in all cases but hydrogen? Moseley had a firm conception of its meaning in relation to the new theory of atomic structure that was being developed by Rutherford and by Niels Bohr (among others). Tragically, Moseley became one of the many millions of casualties of World War I, dying at Gallipoli in Turkey at the age of twenty-seven.

More Radiations: Black Light, Becquerel Rays, and N Rays

X rays were not the only form of radiation to be reported during this era. Their detection immediately led to other announcements of discoveries, both real and spurious. In late January 1896 members of the Paris Academy of Sciences viewed some of Röntgen's X-ray photographs. One of those present was Henri Becquerel.

Like his father and his grandfather, both distinguished physicists, Henri Becquerel's scientific researches included studies of phosphorescence and fluorescence. When Henri Poincaré noted at the academy's meeting of 20 January that it was the phosphorescent point of a cathode ray tube that emits X rays, Becquerel determined to see whether naturally phosphorescent minerals might produce X rays or other radiations that had so far gone unnoticed.

In late February he reported to the academy that he had found that the uranium salt uranyl-potassium sulfate, which becomes phosphorescent after exposure to sunlight, had produced its silhouetted image on a photographic plate that had been wrapped in thick black paper. More amazing were the results Becquerel reported in early March. After several cloudy days during which Becquerel had placed the uranium salt and his wrapped photographic plates in a drawer, he pulled them out and discovered that the plate registered an image as intensely as if the mineral had been phosphorescent. What were these invisible radiations emanating from a natural mineral salt?

Nor was Becquerel's the only such report before the academy. Another experimenter, the popular-science writer Gustave LeBon (1841–1931), announced in late January 1896 the discovery of new radiations he called "black light." Working in his private laboratory, he claimed to have found that ordinary light from an oil lamp produces radiations that affect a photographic plate inside a metal box. Like X rays, he said, these radiations were not deviated in a magnetic field; but, unlike X rays, they discharged the leaves of an electroscope. Several investigators claimed to confirm LeBon's results during February and March, although Poincaré, Gabriel Lippmann, and Auguste Lumiére (1862–1954), who was an expert and entrepreneur in photography, expressed various criticisms.

During the year 1896 the Paris academy heard approximately one hundred papers on X rays, fourteen on black light, and three on Becquerel's rays. By the end of 1897 LeBon's rays were explained away as chemical effects in photographic emulsions and as heat effects in the infrared part of the radiation spectrum. Declining an invitation to demonstrate his experiments as a guest at the academy, LeBon began publishing a revised claim that black light was a stream of electrical particles ejected by ordinary matter and moving at very slow velocities. This claim turned out to be a misunderstanding of well-known electrical phenomena, including the photoelectric effect.

Nonetheless, LeBon reached beyond the scientific community to a broader popular audience with his 1905 book called *The Evolution of Matter.* All matter, he wrote, is dissociating in greater or lesser degrees, continuously emitting radiations such as Röntgen's rays and Becquerel's rays as an effect of the "dematerialization" and "materialization" of matter back and forth from the cosmic ether. Physicists and chemists by and large ignored LeBon's new publications.

Meanwhile, Becquerel and other investigators established during 1896 and 1897 that the natural uranium rays are bent in a magnetic field (and thus they are charged) and that they create charges or ions in gases. These properties provided scientists with quantifiable methods for the detection and further study of the uranium rays with laboratory instruments. Among the many scientists who turned their attention in the late 1890s to the uranium radiations were two advanced physics students: Marie Skłodowska Curie, a Polish emigré in Paris, and Ernest Rutherford, a New Zealand scholarship student in Cambridge.

Marie Skłodowska came to Paris to join her sister Bronia, who was one of the first generation of women permitted to attend lectures and receive a medical degree at the University of Paris. In her turn, Marie Skłodowska passed examinations with high marks in physics and mathematics in 1893 and 1894, and she married the French physicist Pierre Curie in 1895. The two were part of an intimate circle of scientific friends that included Jean Perrin, Paul Langevin, Urbain, Émile Borel (1891–1956) and Borel's wife, Marguerite, who was the daughter of mathematican Paul Appell and wrote novels under the pseudonym Camille Marbo.

After passing the *agrégation,* which was the qualifying examination for teaching in secondary schools and colleges, Marie Curie joined the faculty at the École Normale Supérieure at Sèvres, a teachers' preparatory institution for women in the Parisian suburbs. In the meantime she had embarked on a doctoral thesis.

After making sure that Becquerel had no objection to her investigating the new uranium rays, Marie Curie in 1897 began a program of studying these radiations by measuring their ionization of the air with a very precise piezoelectric quartz electrometer invented by her husband and his brother Jacques Curie. She gathered a great variety of mineral samples in order to examine whether any minerals other than uranium salts emitted radiations, testing the hypothesis of the widespread emission of radiations popularized by LeBon and studied by more mainstream scientists including Robert John Strutt, the fourth baron Rayleigh (1875–1947), J. J. Thomson, and Julius Elster (1854–1920) and Hans Geitel (1855–1923).

Curie established that the intensity of uranium rays is proportional to the amount of uranium present. She soon found that thorium oxide, chalcolyte, and pitchblende not only emit radiations but are considerably more active than uranium salt. In 1898 she began using the term "radioactivity" to describe this "radiance." In what was meant to be short-term help, she enlisted the aid of her husband and the chemist Gustave Bémont to help her identify the source of radiant activity in pitchblende ore.

In 1898 they announced the discovery of "polonium," which was identified in sulfides, and of "radium," which was associated with barium compounds. These new elements were, respectively, 400 and 900 times more radiant-active than uranium. The project took longer than anticipated, and it was four more years before Marie Curie and her collaborators isolated 0.1 decigram of pure radium chloride from several tons of pitchblende. Along the way, in 1899, André-Louis Debierne (1874–1949), who like Bémont was an expert in spectroscopic analysis, discovered actininium. Pierre Curie was in charge of physical measurements of radiation, while Marie carried out chemical analysis and separations.

During this time Pierre Curie, Bémont, and Debierne were all teachers or researchers at the École Municipale de Physique et de Chimie of Paris; they did their work in a shedlike building adjacent to the school. There were career setbacks as well as personal joys in these years, with Pierre Curie losing out to his and Marie's friend Perrin in competition for a position in physical chemistry at the Sorbonne. Pierre Curie then failed to get a chair in mineralogy in Paris after turning down an offer at the University of Geneva. The Curies' daughters Irène and Eve were born in 1897 and 1904.

Marie Curie defended her thesis successfully in 1902, the year before the Curies and Becquerel jointly were awarded the third Nobel Prize in Physics. By 1904 Pierre Curie had been given a newly designated chair at the Sorbonne with a laboratory to be directed by Marie Curie. They had become celebrities not only because of their remarkable work on radioactivity but because of their status as a French-and-Polish husband-and-wife scientific team. Journalists reported conversations between baby Irène and her nurse, and newspapers carried accounts of the black-and-white cat who lived in the Curies' apartment. Photographers and journalists trailed the family everywhere for a few months; then, as the story grew old, they disappeared, allowing the Curies to resume their usual routines—for a time.

In the spring of 1906 tragedy struck. While walking across the rue Dauphine, Pierre was hit and killed by a horse-drawn cart. Marie was awarded a widow's pension by the state, but she immediately refused it and accepted the offer made by the University of Paris sciences faculty to take charge of her husband's physics course at the Sorbonne. In 1908 she became the first woman professor in the university's history.

In 1914, three years after she was awarded the 1911 Nobel Prize in Chemistry, Marie Curie became head of the newly established Radium Institute, located on the street named rue Pierre Curie. She, Linus Pauling, John Bardeen, and Frederick Sanger remain the only people ever awarded two Nobel Prizes.

Marie Curie's life and career provided many firsts for women scientists. The occasion of her inaugural lecture in physics, on 5 November 1906, provided one of the highlights of the Paris autumn social season. By midday a crowd had gathered to compete for seats for the 1:30 P.M. lecture, which was normally attended only by students. Professors, journalists, dandies, and ladies—including the Countess Greffulhe (immortalized by Proust in *A Remembrance of Things Past*)—

were present. One newspaper reported Madame Curie's straightforward lecture on electricity and matter as "a victory for feminism."[2]

The career in radioactivity of Ernest Rutherford (1871–1937) was considerably calmer, but no less illustrious, than that of his colleague across the Channel. Educated at Nelson College and Canterbury College in New Zealand, he received one of the Exhibition Scholarships reserved for British subjects who wanted to carry out advanced studies at universities other than their own. Arriving in J. J. Thomson's Cavendish Laboratory in Cambridge in 1895, Rutherford set to work using the experimental knowledge of magnetism and electric waves that had won him the scholarship.

Following news of Röntgen's and Becquerel's discoveries in early 1896, Thomson set Rutherford and others in his laboratory staff to work on X rays and uranium rays. Given Thomson's long-standing interest in ionization effects in solutions, gases, and vacuum discharge tubes, it is hardly surprising that Rutherford found himself investigating ionization produced by uranium.

While studying the penetrating power of uranium rays through thin metal foils, Rutherford, like the Curies, distinguished two kinds of radiation within the uranium-ray bundle, one species that is very readily absorbed and another that is 100 times more penetrating. For convenience, he called these "alpha" and "beta." By 1900 Paul Villard (1860–1934), working with the Curies in France, discerned a third, "gamma," radiation that was more penetrating than the other two but, unlike them, carried no charge and behaved like X rays. It was now agreed that the beta radiations carry negative charge and are identical to the "electron" defined by Pieter Zeeman in 1896 (see chapter 3) and the "corpuscle" described by J. J. Thomson in 1897 (see below).

During 1902 and 1903 Rutherford and his colleagues established that alpha radiation consists of charged particles that can be bent in a magnetic or electric field and that the value of the mass-to-charge ratio suggests that the alpha "ray" is either a helium ion with a double positive charge or a hydrogen molecule with a single positive charge. Helium had been identified spectroscopically in the sun by J. Norman Lockyer in 1868 but only discovered on earth by William Ramsay in 1895 (in the uranium mineral cleveite).

Rutherford, working at McGill University in Montreal from 1898 to 1907, observed early in his work with Frederick Soddy (1877–1956) that helium was regularly associated with radioactive materials. In 1903 Soddy, in collaboration with William Ramsay

(1852–1916) at University College, London, used Ramsay's sample of radium bromide to confirm spectroscopically that helium is a product of the salt. During 1907 and 1908, Rutherford and his Montreal colleagues collected enough alpha particles from a radium sample to observe the familiar spectral lines identified with helium in the sun.

Bound up with the problem of characterizing the properties of the radiations of naturally radioactive materials was the challenge of understanding the mechanism by which these radiations appear. Early on, in her first paper of 1898, Marie Curie proposed the hypothesis of an external source, or trigger, for the radiations: "One might imagine that all of space is constantly traversed by rays similar to Röntgen rays only much more penetrating and being able to be absorbed only by certain elements with large atomic weight, such as uranium and thorium."[3]

The Curies investigated possible diurnal variations in uranium activity that would show a link to the sun, while in Germany Julius Elster and Hans Geitel investigated whether uranium activity decreased in a vacuum or deep underground—for which experiment they took a uranium sample into a mine 300 meters under the Harz Mountains.

The energy released by a sample of radium was measured precisely in 1903 by Pierre Curie and Albert Laborde, who found that 1 gram of radium can heat 1.33 grams of water from the melting point to the boiling point in 1 hour. The elderly William Thomson responded to this extraordinary experiment by declaring that such energy could not originate inside the atom, and he strongly proposed that etherial waves must somehow supply energy to radium.

Marie Curie also engaged the possibility of an internal atomic source of energy for the radiations, suggesting in another paper, of 1899, that "The radiation [may be] an emission of matter accompanied by a loss of weight of the radioactive substance."[4] This was the hypothesis favored by J. J. Thomson, largely because of his theoretical model of the atom, and it came to be the hypothesis favored by Rutherford and Soddy as the best explanation for the Curies' observation of the radioactivity that was temporarily induced on objects near a uranium or radium source.

Shortly after Rutherford moved to McGill University in 1898, R. B. Owens found that the radioactivity of thorium varied somewhat unpredictably when there were drafts of air. By blowing air over the surface of thorium, Rutherford and Owens were able to collect a gas that they called "thorium emanation." They used electrodes and an electrometer to measure the ionization caused by the gas and found that the radioactivity of the gas decreased as a sample moved through a tube, falling to half its original value in about a minute.

Soddy had joined the chemistry department at McGill in 1900. He began working with Rutherford, who badly needed chemical analysis of the thorium emanation, which resisted reactions with all chemical reagents. When Sir William Macdonald, the millionaire benefactor of McGill University, purchased a liquid air machine for the laboratory, Rutherford and Soddy successfully condensed the emanation.

Complicating the problem was their recognition that more than half the radioactivity of thorium and the entire production of thorium emanation was caused by some other radioactive substance (thorium "X") that could be concentrated in a solution of thorium nitrate, from which a no longer very active thorium hydroxide precipitated. Within a few weeks the original thorium sample recovered its full radioactivity. The conclusion was that thorium produces thorium X, which in turn produces thorium emanation.

In 1902 they identified thorium emanation as a member of the noble gas family (helium, neon, argon, krypton, xenon), eventually calling it radon. Thorium X also seemed to be a different chemical element than thorium, appearing to be a form of radium (what Soddy in 1913 was to christen an "isotope" of radium) with the same chemical properties as radium but a different atomic weight.

Rutherford and Soddy needed a theory that would explain the production of new elements in the radioactive emanations and precipitates, as well as the characteristics of the alpha, beta, and gamma radiations, including the 10,000 mile-per-second velocity of alpha particles. They began to think that atoms of thorium must be spontaneously exploding, hurling out bits of their interiors with enormous energy and leaving behind new and different atoms.

If Röntgen had feared ridicule in 1895, Soddy and Rutherford similarly had much to fear in announcing their new alchemy of transmutations. This they did in a paper entitled "Radioactive Change" published in the leading British physics journal *The Philosophical Magazine* in 1903. Typical was the reaction of British chemist Henry Edward Armstrong. Why, asked Armstrong sarcastically, would atoms find themselves gripped by such "an incurable suicidal mania?"[5]

To make matters more complicated, at this very time when the properties of X rays and radioactivity were under intense investigation, the French physicist René Blondlot (1849–1930) announced what he took to be the discovery of another new radiation. Blondlot, a distinguished professor at the University of Nancy in Lorraine in eastern France, named the radiation "N" to commemorate his city, his university, and Nicolas François, the seventeenth-century regent of Lorraine.

Dozens of people published corroboration of the existence of N rays in 1903, although a larger number could not duplicate Blondlot's results. Originally searching for evidence that X rays are emitted from an electrical discharge tube in the form of a plane-polarized wave, Blondlot reported that he had found a new radiation associated with electrical discharge, but one that, unlike X rays, could be refracted by a quartz prism. As a method of detection, he used changes in brightness of a very weak and short spark.

Astoundingly, Blondlot soon multipled the sources of N rays and their means of detection. He reported that the rays were emitted by both gas and incandescent lamps, by the sun, and by tempered steel. He further claimed that the humoral fluid of the eye stores N rays and reemits them. Even more spectacularly, some of Blondlot's colleagues in the science and medical faculties at Nancy reported that N rays were emitted by animals' nervous systems, including their brains. A new method of detection involved the brightening of phosphorescent spots or of a phosphorescent thread when hit with the N rays; another was the brightening of an electronic spark's impression on a photographic plate if the spark was photographed while illuminated by N rays.

Throughout 1903 and 1904, however, objections continued to be made to the reports confirming the discovery. Many observers perceived no difference in intensity of sparks or spots with or without N-ray illumination. Lummer reported that he and Heinrich Rubens (1865–1922) duplicated the reported "discovery" simply by standing in a dark room, gazing at a screen coated with sulfide spots, and then looking slightly away from the screen so that the peripheral vision of the rods of the retina came into action. Critics demanded that the photographic method should be automated so that no unconscious bias entered the manual procedure of moving the spark device at equal time intervals, with Perrin and Langevin among the few French physicists who strongly questioned Blondlot's claims.

German scientists, some of whom had intially questioned the validity of Becquerel's first report on uranium rays, embroiled the American physicist Robert Wood (1868–1955) in the controversy on the occasion of the late-summer meeting of the British Association for the Advancement of Science in Cambridge in 1904. They persuaded Wood to visit Blondlot's laboratory. Within a few days Wood penned a devastating letter to the widely read journal *Nature* claiming, among other things, that he had clandestinely removed the aluminum prism from Blondlot's apparatus with no effect on Blondlot's results. N rays soon disappeared from the scientific literature, like black light before them.

It was in this chaotic atmosphere of radiation research that Rutherford had been presented the Rumford Medal of the Royal Society and the Curies were awarded the Davy Medal for their work on radioactivity. In the late fall of 1903 Rutherford described his and Soddy's disintegration theory to a London Royal Society audience that included Pierre Curie, who was accepting the Davy Medal on behalf of himself and his wife.

In September of 1904 Rutherford again presented the disintegration theory at the St. Louis Congress of Arts and Sciences, a meeting that included Poincaré and Langevin among its members. Defending the theory of the emission of energy and the "succession" of elements by radioactive transmutation, Rutherford said,

> On this view, there is at any time present in a radioactive body a proportion of the original matter which is unchanged and the products of the part which has undergone change. In the case of a slowly changing substance like radium, this point of view is in agreement with the observed fact that the spectrum of radium remains unchanged with its age.[6]

Exactly why a small fraction of atoms within a sample break up in unit time was unexplained. But, argued Rutherford, the notion that the atom is unstable is not foolish:

> If the atom is supposed to consist of electrons or charged bodies in rapid motion, it tends to radiate energy in the form of electromagnetic waves. . . . According to present views, it is not such a matter of surprise that atoms do break up as that atoms are so stable as they appear to be. This question of the causes of disintegration is fundamental, and no adequate explanation has yet been put forward.[7]

This was in 1904. What was the theoretical model of the atom in 1904?

J. J. Thomson and the Electronic Theory of Matter

During the period from 1895 to 1904 the most popular models for the structure of the atom changed from highly mathematical representations of vortices in the ether to strongly visual images of charged ions or "electrons" in a state of equilibrium. J. J. Thomson was one of the principal architects of mechanical models of the atom in both of these traditions and, indeed, he continued to try to reconcile them through the 1920s.

Thomson took an engineering degree at Owens College in his hometown of Manchester before studying at Trinity College in Cambridge, where he was to become the third director of the Cavendish Laboratory from 1884 to 1918. Thomson's selection for the directorship came as a great surprise to most members of the British physics community both because of Thomson's youth (he was only twenty-eight) and because of his identification with mathematical, not experimental, physics. One of the top contenders, Richard T. Glazebrook, finally recovered from his disappointment and penned Thomson a letter of congratulations with apologies for his belatedness in writing because "the news of your election was too great a surprise to permit me to do so."[8]

Thomson was well known in Cambridge for winning the university's Adams Prize in 1882 with his mathematical essay on the contest's subject of "a general interpretation of the action upon each other of two closed vortices in a perfect incompressible fluid."[9] In the essay Thomson extended his mathematical treatment of mutual attraction of vortices to the chemical combination of atoms forming molecules such as diatomic hydrogen or hydrogen chloride. The following year he applied the vortex-atom theory of gases to explain phenomena of electric discharges in gases, and he became increasingly interested in phenomena associated with the Crookes discharge tube (see chapter 3).

Thomson's widely read book *Recent Researches in Electricity and Magnetism* (1893) had a huge impact on young British scientists such as Ernest Rutherford, John S. E. Townsend (1868–1957) and John A. McClelland (1870–1920), who determined to take their Exhibition Scholarships to the University of Cambridge, which in 1895 admitted non-Cambridge graduates to postgraduate work for the first time.

The arrival of a string of advanced physics students at the time of the discovery of X rays transformed the Cavendish Laboratory. After the mid-1890s, Thomson's investigations and the projects that he assigned to his staff and advanced students centered squarely on cathode rays and X rays. Thomson aimed to disprove Lenard's 1894 contention that cathode rays are not charged particles, and he wanted to prove that X rays are pulses in the ether produced when fast-moving charged particles from the cathode hit glass or metal in the vacuum tube.

An initial series of investigations on the velocity of the cathode rays convinced Thomson that they traveled at a velocity 1,000 times slower than the velocity of light and that he was correct in arguing that they could not be electromagnetic radiations. (The velocity figure turned out to be too low, as he realized by 1897.)

The year 1897 is often given as the year of the "discovery" of the electron. In May 1897 Thomson delivered a Friday Evening Discourse at the Royal Institution; the topic of his lecture was also the subject of his paper on cathode "rays" that was to appear in the October 1897 *Philosophical Magazine.* This experimental paper was largely responsible for Thomson's change in reputation from a "mathematical" to an "experimental" physicist.

Thomson's paper is well known for his calculation of the mass-to-charge ratio for the cathode-ray corpuscles, although the paper does not mention the similar value found a year earlier by Lorentz and Zeeman for the "ion" in the atom reponsible for the widening of spectral lines in a magnetic field. Thomson's paper argues mainly against Lenard's and other "German physicists'" results, using as its starting point an 1895 experiment by Perrin that established that the charge of a Faraday cylinder is always negative when a pencil of cathode rays falls into it, but that the charge disappears when the cathode rays are deviated away from the cylinder by a magnetic field.

To dispel any doubts that the negative electric current always and only follows the exact path of the rays, Thomson modified Perrin's apparatus in order to bend the rays around corners from their origin at the cathode to the cage attached to an electrometer. He showed that the electrometer registers an electric charge only when rays are directed into the Faraday cylinder.

Thomson further demonstrated that Hertz's failure to detect cathode-ray deflection between electrostatically charged plates was the result of an inadequate vacuum in the German physicist's discharge tube, so that positive- and negative-charged ions from gas remaining in the tube coated the sides of the tube when the plates were turned on.

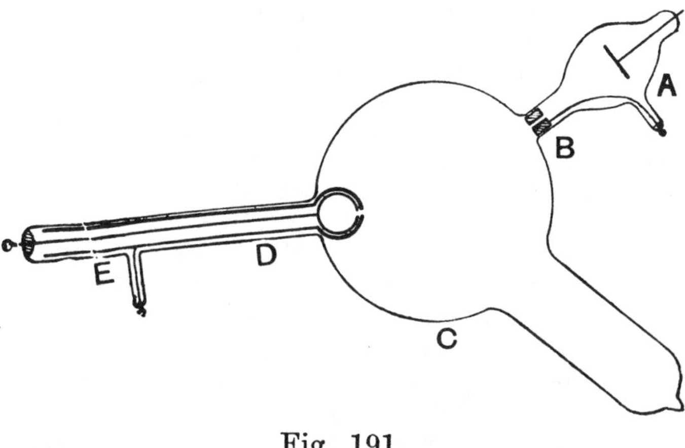

Fig. 191.

J. J. Thomson's device for demonstrating that cathode rays carry an electric charge. An electrometer connected to inner cylinder E registers a substantial negative electric charge only when the rays starting from cathode A are deflected into the inner cylinder E by a magnet. From J. J. Thomson, *Conduction of Electricity through Gases,* 2d edition (Cambridge: Cambridge University Press, 1906). *Courtesy of Cambridge University Press.*

Thomson's calculation of the mass-to-charge ratio of the cathode-ray corpuscles came from a clever use of magnetic and electric fields for deviating the path of the cathode rays, so that the *m* to *e* ratio became the unknown quantity deduced after precisely balancing the two fields so that there was no angular deflection of the cathode rays.

Using this and other methods Thomson placed the corpuscles' velocity at 10^9 centimeters per second rather than at his earlier calculation of 10^7 centimeters per second. The figure for m/e was found to be independent of the nature of the gas in the discharge tube. Its value was 10^{-7} in comparison with the value 10^{-4} for m/e of the hydrogen ion in electrolysis. Given a choice between the hypotheses that the mass of the corpuscles was 1,000 times smaller than a hydrogen ion *or* that the charge of the corpuscles was 1,000 times larger than a hydrogen ion, Thomson (like Lorentz) did not hesitate:

> Thus on this view we have in the cathode rays matter in a new state, a state in which the subdivision of matter is carried very much further than in the ordinary gaseous state: a state in which all matter—that is, matter derived from different sources such as hydrogen, oxygen, &c.—is one and the same kind; this matter being the substance from which all the chemical elements are built up.[10]

Thomson proposed a model of the atom made up of large numbers of "primordial atoms." Because the equations for stability of such a collection of particles would be so complex, a mechanical model rather than a mathematical investigation must do, he told his Royal Institution audience and his *Philosophical Magazine* readers. Chemistry and physics came together in Thomson's notion of an atom model that might explain the periodic properties of the chemical elements. Atoms, he hypothesized, may simply be complicated analogs of floating magnets, in which the magnets arrange themselves in equilibrium under the combined conditions of their mutual repulsions and the central attraction caused by the pole of a large magnet placed above the floating magnets.

It had been shown that when the number of floating magnets was between three and five, they arranged themselves at the corners of a regular polygon: three in a triangle, four in a square, five in a pentagon. Yet when the number of magnets exceeded five, they broke up into two groups, with one magnet alone in the middle surrounded by five arranged at the corners of a pentagon. For eight, there were two inside and six outside. Eighteen floating magnets broke into three systems. If atomic corpuscles or "primordial atoms" arranged themselves into analogous structures, the periodicity of properties of the chemical elements might (ingeniously) be explained by the recurrence of these internal and external patterns.

From this model, Thomson moved on to other models, struggling to figure out just how many hundreds or thousands of these tiny corpuscles might be needed to explain atomic properties if atomic weight were due only to the weight of the corpuscles. By 1899 Thomson, Townsend, and H. A. Wilson had measured the corpuscle's charge directly, using the technique of calculating individual charge from the total charge on fog condensed on ions in the absence of dust particles. This technique and its apparatus were developed by the physics demonstrator and tutor C. T. R. Wilson (1869–1959). The mass of the corpuscle was now thought to be even smaller than first estimated: almost 2,000 times smaller than a hydrogen atom, not just 1,000 times smaller.

By 1904 Thomson inclined toward a model of the structure of the atom that he knew was hopelessly complicated. In his Yale lectures in 1903, Thomson proposed that the simplest atom of hydrogen consists of a spherical shell over which positive charge is uniformly distributed. Inside the shell, distributed in concentric spheres like currants inside an English Christmas plum pudding, are approximately 2,000 corpuscles that give the hydrogen atom its weight.

By 1906 Thomson was using a hydrogen-atom model of a single electron embedded in a positively charged jelly. The need for thousands of electrons had been destroyed by experiments conducted by Charles G. Barkla (1877–1944) and C. A. Sadler at Liverpool. They used X-ray scattering to determine that the number of electrons in the carbon atom is only about half the carbon's atomic weight—in other words, six electrons. At about this time Barkla also demonstrated that X rays can be plane-polarized, undermining Thomson's confidence in the pulse theory of X rays.

By this time, the Japanese physicist Hantaro Nagaoka (1865–1950) and Perrin had independently proposed alternative atomic models. Nagaoka was well versed in current European physics. He had studied in Vienna and Berlin, giving a paper in Paris at the 1900 International Physics Congress before returning to the University of Tokyo to set up a physics laboratory modeled on the best ones on the European continent.

Nagaoka's atomic model of 1904 was a sort of inverse plum pudding: a halo of electrons surrounding a positively charged central sphere. Perrin's earlier model of 1901 had gone largely unnoticed because it had appeared in a French popular-science journal. Perrin argued that, since atoms can be ionized into a small piece that is negatively charged and a larger piece that is positively charged, each atom is constituted

> by one or several masses very strongly charged with positive electricity, in the manner of positive suns whose charge will be very superior to that of a corpuscule, and on the other hand, by a multitude of corpuscules, in the manner of small negative planets, the ensemble of their masses gravitating under the action of electrical forces.[11]

In 1908 Johannes Stark (1874–1957) proposed yet another model of the atom: a surface consisting of spheres or spherical zones of positive electricity with small, pointlike negative electrons lying between or above the concentric spheres.

To all these models, the same objection was made that Rutherford noted at St. Louis in 1904. If electrons were in motion within the sphere of the atom, it was no surprise that atoms would lose energy in the form of electromagnetic radiation. But why were so few atoms unstable, and why were most atoms stable at all?

Rutherford, Bohr, and the Quantized Nuclear Atom

In 1908 Rutherford took over the physics laboratory at Manchester, where he remained director until returning to Cambridge in 1919 to succeed J. J. Thomson. Rutherford's Manchester laboratory, with a staff of some fifteen to twenty people, quickly became the world's preeminent facility for research on radioactivity.

During these years Hans Geiger, B. B. Boltwood, Kasimir Fajans, Georg Hevesy, E. N. da C. Andrade, James Chadwick, George C. Darwin, Ernest Marsden, Harry Moseley, and Niels Bohr were among the many physicists who joined Rutherford's research group. Their work in radioactivity was done with a small quantity of radium bromide (350 milligrams) lent to Rutherford and later purchased from the Vienna Academy of Sciences. The samples that the academy lent to both Rutherford and William Ramsay came from its source in the Joachimstal (Jáchymov) mines in what is now the Czech Republic.

One of the principal projects in Rutherford's laboratory was the testing of hypotheses about the distribution of electrical charge within the atom. The testing consisted of studying the paths of alpha or beta radiations directed against metal targets; researchers found that the scattering was more marked for the beta than the alpha particles because of the beta particles' smaller momentum and energy. The overall scattering was assumed to be the result of a multitude of small scatterings within the atom's structure.

Because of the assumption that the mass of the atom, as well as its negative and positive charge, is distributed fairly uniformly throughout the atom's sphere, the relatively heavy alpha particles were expected to move in a more or less straight line through thin metal foil. If, however, Hantaro Nagaoka's model which placed the positive charge within a sphere inside an electron halo—were correct, an alpha particle should experience electrostatic repulsion near

Staff and research students in the Chemistry Department at the University of Manchester, 1907–1908. Many chemists attended the weekly Friday physics colloquia organized by Ernest Rutherford, who received the Nobel Prize in Chemistry in 1908. Among future Nobel laureates in chemistry were Walter Norman Haworth in 1937 (back row, far left) and Robert Robinson in 1947 (middle row, far left). Others pictured include Jocelyn F. Thorpe (front row, third from left) and Ida Smedley [Maclean] (front row, third from right). Of William Henry Perkin, Jr. (front row, fifth from right), Rutherford is reported to have said in a lecture: "the nucleus is a round, hard object—just like Professor Perkins's head." From G. N. Burkhardt, "Schools of Chemistry in Great Britain and Ireland-XII: The University of Manchester (Faculty of Science)," *Journal of the Royal Institute of Chemistry* 78 (1954), 448–460; illustration facing p. 455. *Reproduced Courtesy of the Library and Information Centre, Royal Society of Chemistry.*

the positive sphere and suffer more marked deflection than it would according to the "plum pudding" model.

The method of following the alpha particle was extraordinarily tedious. In a dark room one or two observers used a movable, low-powered microscope to count flashes on a zinc sulfide screen as the charged particles hit it one by one. At each scintillation the observer pressed a key that recorded his observation. Then the different counts were correlated to come up with a reliable number.

One of the lasting contributions to the science of radioactivity was the invention by Hans Geiger (1882–1945) of the "Geiger counter," which he perfected with Walther Müller after returning to Germany. It replaced the visual method of detection, which was subjective and often difficult to duplicate in other laboratories, with an automated electrical device that registered the ionization caused by each alpha particle. If Blondlot had been able to come up with an analogous device for N rays, his discovery might have been revived.

Ernest Marsden (1889–1970) came from New Zealand to work with Rutherford in 1909. He soon reported to Rutherford that he had observed occasional deflections of the alpha particles at very large angles. Working together, Marsden and Geiger within the next couple of years established that a small fraction of particles in the alpha-ray beam, about one in twenty thousand, was deflected at an average angle of 90^0 instead of passing through a sheet of gold foil about 0.00004 centimeters thick, which, Rutherford noted, was equivalent in stopping power to 1.6 millimeters of air. As Rutherford reminisced in the 1930s, "It was about as credible as if you had fired a 15-inch shell at a piece of tissue paper and it came back and hit you."[12] The probability that this strong deflection could occur as a result of multiple scatterings was vanishingly small.

In 1911 Rutherford postulated that an atom contains a positive charge *Ne* ($e = 4.65 \times 10^{-10}$ e.s.u.) at its center and is surrounded by negative electricity *Ne* uniformly distributed within a sphere of radius *R*. From this hypothesis he deduced equations for the deflection of a positive electrified particle traveling close to the center of the atom, supposing its velocity not to be appreciably changed and its path to fall under the influence of a repulsive force varying inversely as the square of the distance. A comparison of the theory with experiments carried out by Geiger and Marsden was confirming. Experiments by James Arnold Crowther (1883–1950) on beta-ray scattering indicated that the central charge is approximately proportional to atomic weight.

Rutherford's preoccupation with the mechanism of radioactivity in relation to atomic structure appeared at the conclusion of his 1911 paper, where he noted that the approximate value for the central charge of an atom of gold "is about that to be expected if the atom of gold consisted of 49 atoms of helium, each carrying a charge $2e$."[13] The high velocity of expulsion of alpha particles from radioactive atoms could be explained by the helium ion's acquiring a large velocity while moving away from the core of the atom through the electric field. Rutherford's notion of this nuclear core, shared by most physicists, was that it contained positively charged heavy matter and negatively charged light electrons, with additional electrons, numbering approximately half the atomic weight, arranged in rings or spheres at some distance from the nucleus.

In the fall of 1911 Niels Bohr (1885–1962) arrived in Manchester to study radioactivity measuring techniques after a short stay with J. J. Thomson at the Cavendish Laboratory. Bohr had written his doctoral dissertation in Copenhagen on the electronic theory of metals and he shared the interest of the Danish and German scientific community in the quantum theory, which was a focus of discussion at the Solvay Physics Conference in Brussels that same autumn. Bohr's interest in the subject extended to wondering whether the energy of electromagnetic radiations emitted by the atom's electrons might not best be described by Planck's quantum formulation for black-body radiation and Einstein's quantum formulation for light.

This was not an idea that much interested Thomson, who continued to use late nineteenth-century formulations of etherial tubes of force in his descriptions of the energy of the electron. At Manchester, Bohr found a more congenial intellectual atmosphere for developing his hypotheses, and by the summer of 1912 he prepared a memorandum for Rutherford on the subject of atomic structure. The memorandum showed the considerable influence of Thomson's way of thinking about the relationship between atomic structure and chemical periodicity and demonstrated Bohr's knowledge of chemistry, which had been augmented in long conversations at Manchester with Georg Hevesy (1885–1966). Bohr was to return to Denmark during 1912 and 1913, where he married, but he came back to Manchester in the position of Schuster Reader during the years 1914 to 1916.

Bohr's memorandum, which later became a section of a three-part paper "On the Constitution of Atoms and Molecules," concerned itself with the relationship among electron rings, the periodicity of the chemical elements, and the problem of chemical binding. Bohr used the concept of affinity "work," calculating the quantity of work (energy) needed to remove an electron from diatomic hydrogen, as well as the heat released when two atoms of hydrogen form a molecule of hydrogen. His paper confirmed from a theoretical framework the well-known facts that diatomic helium does not exist and that diatomic oxygen does not dissociate into ions.

Bohr modeled the simple diatomic molecule as a girdle of electrons rotating in a circle at a right angle to the axis connecting two atoms. He attributed the chemical properties of an atom to electrons well outside the nucleus of the atom and any radioactive properties to the nucleus. He substituted the language of electron "shells" and "subshells" for the teminology of "orbits" or "spheres." But what eluded Bohr was a way of determinining the distance of the outer electrons from the nucleus, even in the simplest case of the hydrogen atom with its one electron.

Early in 1913 Bohr's Danish friend Hans Marius Hansen suggested to Bohr that his model really should include information about spectra. Hansen recommended that Bohr look at the work on hydrogen spectra by J. J. Balmer (1825–1898) as revised by Robert Rydberg (1854–1919) of the University of Lund, Sweden. Bohr later recalled, "As soon as I saw Balmer's formula the whole thing was immediately clear to me."[14]

The spectrum of hydrogen is not continuous like the spectrum of white light but consists of a few bright, narrow lines corresponding to a few discrete wavelengths. Rydberg's formulation, introducing a constant that became known as the Rydberg constant was

$$\frac{1}{\lambda} = R\left(\frac{1}{2^2} - \frac{1}{n^2}\right)$$

The generalization of this formula, expressed in frequency rather than wavelength, is

$$\nu = R\left(\frac{1}{n_1^{\,2}} - \frac{1}{n_2^{\,2}}\right)$$

where n_1 and n_2 are positive integral numbers and $n_1 < n_2$.

Three different series of lines were known in 1913, corresponding to $n_1 = 2$ (Balmer's series in the visible range), $n_1 = 1$ (Theodore Lyman's series in the far ultraviolet region, worked out between 1906 and 1914), and $n_1 = 3$ (Friedrich Paschen's two lines in the infrared, discovered in 1908). What Bohr saw in the formula was a way of using a simple law of radiation to describe the energy of the orbital electron and to fix the radius of the orbit.

Bohr's model used two apparently contradictory laws of physics. On the one hand, he employed the classical description of the energy of a point mass in the tradition of Newtonian mechanics and Coulombian electrostatics. But he also used the formulation of Planck and Einstein for the discontinuous emission of energy in units defined by the quantum of action *h*. The validity of the resulting model was to be judged by its predictive power (e.g., the Lyman and Paschen series for hydrogen), its calculation of a reasonable value for the radius of an atom, its explanation of the physical meaning of the constant *R* in Balmer's formula, its answer to the conundrum of the ordinary atom's stability, and its potential for further development.

Bohr's reasoning, much simplified, was as follows: Take an atom to be formed by a massive center with positive charge attracting an electron. The orbits will be Keplerian ellipses with the nucleus in one focus. This orbit can be approximated by a circle, along which travels the electron with mass *m* and velocity *v* at a radius *r*. Balancing centrifugal force and Coulombic attraction gives

$$\frac{mv^2}{r} = k_e e \cdot \frac{e}{r^2}$$

The speed *v* of the electron can also be expressed as

$$2\pi fr$$

so that, by substitution and rearrangement,

$$\frac{1}{f^2} = T^2 = \left(\frac{4\pi^2 m_e}{k_e e^2} \right) r^3$$

This relationship highlights the classical Keplerian law

$$T^2 \text{ proptl } r^3$$

and the change in period *T*, or frequency, of the electron's orbit with its distance from the nucleus.

What Bohr then did was to state that the electron obeys the quantum law, emitting or absorbing energy in discrete units

$$\Delta E = h\nu$$

In this formulation, however, ν is not the same frequency as f for the orbital revolution of the electron. Energy is not emitted continuously during the electron's orbital motion at frequency f, but only as the electron moves inward from an orbit at some discrete radius to another, smaller orbit. In his final derivation, the frequency ν is expressed as

$$\nu = \frac{2\pi^2 me^4}{h^3}\left(\frac{1}{n_1^2} - \frac{1}{n_2^2}\right)$$

Thus, with $R = 2\pi^2 me^4/h^3$ and $n_1 = 2$, Balmer's formula for the series of hydrogen spectral lines in the visible region is derived.

The quantity h, the quantum of action, is a measure of the product of the mass of the electron by its velocity and by the distance it travels. In formulating the energy of the electron, Bohr used the convention that the energy has zero value at an infinite distance from the center of the atom and a negative value close to the center. Positive energy values result in an unstable and explosive atom.

By the time Bohr completed the paper, positioning his most recent and revolutionary part at the beginning as Part I, he had been in Denmark for some time. He sent a copy of the paper to Rutherford with the request that Rutherford submit it to the *Philosophical Magazine* on Bohr's behalf, Soon after, Bohr found himself on a boat back to England: Rutherford had responded that the paper was too long and too "wild." Meeting with him, Bohr prevailed over Rutherford, who wrote a friend that he never would have expected such obstinacy from the young Danish physicist.[15] Hevesy described Einstein's reaction to Bohr's paper in a letter to Rutherford: "The big eyes of Einstein looked bigger still, and he told me 'Then it is one of the greatest discoveries.'"[16]

It was not until 1920—after the terrible years of World War I—that Bohr met Einstein for the first time. By then, Bohr had become professor of theoretical physics at the University of Copenhagen. In the following year, his new Institute for Theoretical Physics was completed. Until the next world war again disrupted all of Europe, Bohr's institute in Copenhagen was to become a step in the rite of passage for many young theoretical physicists whose ideal intellectual pilgrimage included sojourns in the theoretical institutes of Arnold Sommerfeld in Munich, Max Born in Göttingen, Erwin Schrödinger in Zurich, and Bohr in Copenhagen.

Among Bohr's coworkers in the 1920s were H. A. Kramers and Paul Ehrenfest from the Netherlands, Hevesy from Hungary, P. A. M. Dirac and Nevil F. Mott from Great Britain, Oskar Klein from Sweden, Werner Heisenberg from Germany, Wolfgang Pauli from Austria, George Gamow and L. D. Landau from Russia, Yoshio Nishina from Japan, and John Slater, Robert Oppenheimer, and Harold Urey from the United States. Bohr received the Nobel Prize for Physics in 1922.

From Quantum-Electron to Quantum-Wave Mechanics

For almost thirty-five years after he was appointed to a chair in theoretical physics at the University of Munich, Arnold Sommerfeld (1868–1951) attracted gifted students, including Heisenberg, to his Munich institute. In 1915 Sommerfeld published a book, *Atomic Structure and Spectral Lines,* that was to become a sort of bible of the new physics. Max Born's *Constitution of Matter,* published in 1920, brought Sommerfeld's ideas and formulations to an even wider audience of scientists.

What Sommerfeld did was to study the properties of an elliptical electron orbit, for which Bohr's circular orbit is a special case. Following up on a paper published by Bohr in 1915, Sommerfeld took into account the theory that as the velocity of an electron mass changes with its position in an elliptical orbit, there must be a change in its mass in accordance with the relativistic theories of Lorentz and Einstein. Thus Sommerfeld provided powerful calculations for the effects of relative mass in electron theory, empirically observed in the fine structure of the Balmer spectral lines.

Sommerfeld calculated elliptical electron orbits, defining an "azimuthal" or "second" quantum number in addition to Bohr's "first" quantum number for the radius of the principal axis of orbit (n_r or n). The second quantum number specifies the orbit's eccentricity as an effect of its quantized angular momentum. The orbit of lowest energy is, of course, the circular orbit, and for each integral value of n, the higher the value, the greater the energy.

Sommerfeld further introduced a third quantum number, which specifies quantized orientation in space of the plane of the orbit in a magnetic field. This property of orientation can be related empirically to the splitting of spectral lines in a magnetic field (the "normal Zeeman effect").

Following up on Sommerfeld's approach, Bohr in 1922 outlined an improved general model of the structure of atoms using what he called the *Aufbauprinzip* ("building-up principle") for feeding electrons into an atom's subshells, a principle he had already proposed in 1912. He arranged these subshells into groups of (2), (2,6), (2,6,10), and (2,6,10,14). Correlating the electrons in these subshells with spectral-line data for the so-called strong, principal, diffuse, and fundamental lines, he designated the subshells *s, p, d,* and *f.*

Among the spectral-line problems still puzzling theoretical physicists was the "anomalous Zeeman effect": the splitting of lines in a magnetic field. These lines were deemed anomalous in the context of the theoretical requirement that the electron's angular momentum and direction in space in the presence of a magnetic field must be calculated using integral multipliers of $h/2\pi$. Yet formulas that worked for describing the "anomalous" Zeeman lines were half-quanta formulas. Among those obsessed with this problem was Wolfgang Pauli (1900–1958). When asked one day in Copenhagen why he looked so depressed, he is reported to have complained, "How can one avoid despondency if one thinks of the anomalous Zeeman effect?"[17]

In a paper published in 1925 Pauli conquered his depression by proposing a practical solution: the use of a fourth quantum number, which can have two values. If it is assumed that no two electrons in an atom may occupy the same atomic energy state, meaning that no two electrons can have the same four quantum numbers, then there might be two—but no more than two—low-energy (*s*) electrons for each principal quantum number. Six different *p* electrons would be possible, ten *d,* and fourteen *f.* What bothered Pauli about this principle of "exclusion," however, was that it had no logical or physical basis.

In the late fall of 1925 the Dutch physicists George Uhlenbeck (1900–1988) and Samuel Goudsmit (1902–1978) gave a physical interpretation to Pauli's postulate of a fourth quantum number. The electron, they proposed, may spin in one of two directions. In a given atom, a pair of electrons having three identical quantum-number values must have their spin axes oriented in opposite directions, and, if paired oppositely in a single orbital, they neutralize each other magnetically. This interpretation fit neatly within Bohr's program of finding "correspondence" of the quantum theory with classical mechanics and electromagnetism since the hypothesis of the spinning electron particle was a very classical idea. Bohr's "correspondence" principle was soon to give way, however, to a newer and more radical principle of "complementarity" following novel work by the very young Heisenberg.

During June 1925 Werner Heisenberg (1901–1976) took a brief vacation and sought relief for an attack of hay fever by spending ten days on the tiny resort island of Helgoland off the north German coast. He had been in Göttingen, in close contact with Max Born (1882–1970) and Born's assistant Pascual Jordan (1902–1980), following a fellowship in Copenhagen. As Heisenberg's biographer David Cassidy puts it, Heisenberg "killed off" mechanical orbits that summer, aided and abetted by Born, who realized before Heisenberg that Heisenberg's approach involved the old but little-known abstract mathematics of matrix calculus.[18]

Arranging frequencies or intensities of spectral lines in tabular form, Heisenberg formulated a mathematical theory that used only empirical data. The method labeled every empirical quantity by two indexes, one corresponding to an initial state and the other to the final state of the system under description. As further developed with Born, one index corresponds to the row and the second index to the columm of a square in an array of squares in which the number is located. Insofar as there is an underlying physical model, it is that of an oscillator with momentum p about an equilibrium position q.

The introduction of the new *Matrizenmechanik* was foreign and uncongenial to all but a few theoretical physicists. One to whom it was not uncongenial was Paul Adrien Maurice Dirac (1902–1984), to whom Heisenberg mailed his page proofs at Cambridge. Dirac, independently from Heisenberg and Born, enthusiastically began working out a general, abstract quantum mechanics for the electron.

In the meantime a young prince, Louis de Broglie (1892–1987), a member of a military and diplomatic family with a pedigree going back to the Broglia family in twelfth-century Piedmont, published several articles and a doctoral thesis that provided an astonishing, yet attractive, theory of the electron.

De Broglie's education and situation were unusual among twentieth-century physicists. His elder brother, Maurice de Broglie (1875–1960), inherited the title and responsibilies of the dukedom conferred on François-Marie de Broglie (1671–1745), an army marshal and ambassador to London whose chateau domain in Normandy took the name Broglie in 1742. Louis de Broglie, like his brothers and sisters, carried the title "prince" (or "princess") as the result of a benefice made by the Holy Roman emperor Francis I to the first duke's son Victor François for his military exploits. Maurice de Broglie served in the navy and then abandoned the family's military and diplomatic tradition by taking up science privately. The example of the older brother paved the way for the younger one.

Louis de Broglie's intellectual interests shifted from history and philosophy to the sciences shortly after he passed his *licence* (undergraduate) exams at the University of Paris. The shift became decisive after he read the page proofs for the proceedings of the 1911 Solvay Physics Conference, where his elder brother served as secretary. Thereafter, Louis began spending most of his time studying physics, with special interest in Planck's and Einstein's theories, while experimenting in his brother's private laboratory housed in his residence on the rue Chauteaubriand.

Scientific education and experimentation were interrupted by the war, during which Louis served in the French army in radio communications, based in the Eiffel Tower. In the early 1920s laboratory life on the rue Chateaubriand resumed, including weekly meetings on Wednesdays among Maurice's informal students, collaborators and staff. These men included Alexandre Dauvillier, Jean Thibaud, Jean Jacques Trillat, Louis Leprince-Ringuet, and younger brother Louis. The laboratory specialized in X-ray studies, which were Maurice's longtime interest.

In 1923 Louis published three papers in the *Comptes rendus*, the weekly journal of the Academy of Sciences. The following year he defended a thesis based on these papers before a jury that included the crystallographer Charles Mauguin and the mathematician Elie Cartan, as well as Perrin and Langevin. Meeting the philosopher Léon Brunschvicg in the Latin Quarter after having glanced at the thesis, Langevin commented, "I am taking with me the little brother's thesis. Looks far-fetched to me." A closer reading led Langevin to instruct the thirty-one-year-old "little brother" to mail the thesis to Einstein. Langevin soon had Einstein's opinion: "He has lifted a corner of the great veil."[19]

Among the audience at the public defense was R. J. Van de Graaff (1901–1967), who was studying at the Sorbonne. "Never," he said later, "had so much gone over the heads of so many."[20] Within a year Dirac asked de Broglie for a copy of the thesis. It was studied in Munich by Sommerfeld and in Göttingen by Heisenberg, Born, and Jordan. In March 1926 Erwin Schrödinger wrote to de Broglie to say that he had extended de Broglie's ideas into a new theory of his own.

In his papers and thesis of 1923–1924, de Broglie, reasoning by analogy from the quantum theory of radiation and the photon theory of light, proposed to associate a real physical wave with any free particle (such as an electron). Using relativistic arguments, he derived the value of the wavelength

$$\lambda = \frac{h}{mv}$$

in relation to the quantum of action and the momentum of the electron, thus associating a wavelength directly with a particle for the first time.

He also worked out a theory of interference and diffraction for the electron that could be subjected to experimental verification. In 1925 de Broglie asked Dauvillier to attempt to confirm his predictions for electron diffraction in crystals. Dauvillier, however, had no immediate success. It was not until 1930 that electron diffraction was achieved in France, and by this time it had been accomplished by Clinton J. Davisson (1881–1958) and Lester H. Germer (1896–1971) at the Bell Laboratories in New Jersey and by J. J. Thomson's son George P. Thomson (1892–1975) at Aberdeen, Scotland.

De Broglie's theory became known as the pilot-wave theory of the electron. He compared an observable particle such as the electron to a microscopic granule in contact with a hidden reservoir of activity, in analogy to a pollen granule in Brownian motion (see chapter 4). Two mathematical solutions must be possible for a wave mechanics: one with continually variable amplitudes that describes the statistical aspect of electron movement and a second that uniquely represents the corpuscle at the heart of the wave. De Broglie vigorously defended this two-solution theory through 1927, when he argued in its favor at the Fifth Solvay Physics Congress in Brussels.

Schrödinger, along with Einstein and Lorentz, was sympathetic with de Broglie's attempt to find a realistic and deterministic account of particle-wave duality that would reconcile classical and quantum mechanics. They all wanted to ensure physical objectivity to the electron independent of the observer.

But by the fall of 1928 de Broglie succumbed to the emerging "Copenhagen interpretation" of quantum mechanics. His lectures at the University of Paris sciences faculty began to describe the up-to-date theories of Bohr and Heisenberg and to outline the inadequacies of the pilot wave. Only with the reappearance of the pilot wave in Princeton physicist David Bohm's controversial theory of 1952 did de Broglie's approach come into serious discussion again, principally in order to be rejected once more.

Erwin Schrödinger (1887–1961), who had studied physics and mathematics in Vienna, was teaching theoretical physics at the University of Zurich in 1926 when he published his treatment of the electron based on a reformulation of de Broglie's early work. What Schrödinger did was to apply the well-known Hamiltonian variational principle to the matter-waves that de Broglie associated with electrons, reasoning that the electron can be thought of as an intense concentration of waves in a small space. Schrödinger worked out an equation in a form that was familiar to mathematical physicists studying acoustic or electromagnetic waves:

$$\nabla^2\Psi(x,y,z)+\left(\frac{8\pi^2 m}{h^2}\right)\{E-V(x,y,z)\}\Psi(x,y,z)=0$$

where ∇^2 is the Laplace operator, *m* is the mass of the electron, *h* is Planck's constant, *E* is the total energy of the particle, and *V* is the electron's potential energy in a field of force.

The function ψ was taken to be the amplitude of the electron's matter wave. The solutions of Schrödinger's equation were such that $\int|\psi(x,y,z)|^2$ is finite, and the values of *E*, called "eigenvalues," correspond to the possible energies of an electron in nonradiating states in the potential *V*.[21] That is, the eigenvalues represent the energies of a standing wave in distinct modes of vibration. The size of the (Bohr) electron orbit is conditioned by the requirement that only an integral number of waves may oscillate on the orbit.

The function ψ (or the more general time-dependent wave function Ψ, which is the product of two functions, one involving the time alone and the other the coordinate alone) did not have a straightforward physical meaning, because the electron wave is more abstract than the sound wave or electromagnetic wave, for which the analogous equations determine the pressure or electric-field strength. Since in optics the square of ψ is proportional to the intensity of light, the square of ψ in Schrödinger's equation was taken by him to be a measure of electron intensity, or density, about the atom nucleus.

Theoretical physicists' interest in Schrödinger's work was intense. Bohr invited him to Copenhagen, where they argued over the meaning of ψ^2, as well as over the notion of discontinuities and the hypothesis of quantum "jumps." In July 1926, Wilhelm Wien (1864–1928), now rector at the University of Munich, and his colleague Arnold

Sommerfeld invited Schrödinger to lecture on the new theory. Wien , clearly annoyed at Heisenberg's ardent questioning of Schrödinger about how the new theory could explain elementary quantum phenomena such as Planck's radiation formula and the Compton effect, at one point motioned Heisenberg to sit down and be quiet.[22]

Schrödinger's requirement that a physical theory should be "visualizable" or "graphic" (*anschaulich*) was absolutely unacceptable to Heisenberg, who wrote to Pauli, "The more I think about the physical portion of the Schrödinger theory, the more repulsive I find it."[23] Heisenberg passionately wanted to break with both classical and continuum physics.

Nor was Heisenberg pleased with the interpretation suggested by Born, who returned to Göttingen after lecturing in the United States during 1925 and 1926 and who spoke of the electron in a traditionally realist and visualist way. Schrödinger's equation, Born argued, should be interpreted statistically as the probability of finding electron particles at certain locations in space and time, rather than as the electron density of matter or charge in space. Born's interpretation was meant to be consistent with the results of X-ray scattering in crystals and to fit in with the statistical character of thermodynamics, radioactivity, and Einstein's work (1916) on transition probabilities for electron jumps.

While convincing himself that matrix mechanics and wave mechanics are equivalent mathematically, Heisenberg also examined the meaning of the new quantum theory for the kinds of questions and answers that it is possible to address experimentally. What can we know about the precise or "true" path of an electron in the cloud chamber if the electron can be described by a wave equation? As Heisenberg wrote, "if we want to know of a wave packet both its velocity and its position, what is the best accuracy we can obtain, starting from the principle that only such situations are found in nature as can be represented in the mathematical scheme of quantum mechanics?"[24]

If we want to locate the electron precisely, we must use radiation of very small wavelength and high energy, which, given the photon identity of radiation, collides with the particle and alters its velocity. If, however, we use radiation of a longer wavelength, we can no longer locate the electron precisely although we now do not affect its velocity. So, if Δp is the inescapable error of coordinate measurement and Δq the inescapable error for momentum measurement ($q = mv$), then

$$\Delta p \, \Delta q \geq \frac{h}{2\pi}$$

as deduced from the quantum condition

$$pq - qp = \frac{h}{2\pi i}$$

This was Heisenberg's "uncertainty principle." Along with the relative strengths and weaknesses of the theories of wave and matrix quantum mechanics, it became the subject of intense debate at the September 1927 conference of theoretical physicists at Lake Como in Italy. It was on this occasion that Bohr offered the views that became known as the "Copenhagen interpretation" of quantum mechanics or as Bohr's theory of "complementarity."

Physicists must make a choice. They may choose to interpret quantum-level phenomena either as particle-like causal events in a physics of discontinuities or as wavelike, space-time events in a physics of the continuum. Heisenberg's principle is a statement, Bohr argued, of the way in which the experimenter's choice necessarily introduces a disturbance into the phenomenal situation so that precise knowledge cannot be obtained for that side of the wave-particle duality not chosen.

As mentioned above, this "Copenhagen interpretation," argued again in October 1927 at the fifth Solvay Congress in Brussels, was hotly contested by Schrödinger, Louis de Broglie, and others. Einstein occasioned considerable notice by confronting Bohr with new objections at breakfast. For the rest of his life, Einstein insisted that indeterminism lies in our present methods for knowing the object, not in the object itself.

For the rest of their lives Bohr and Heisenberg disagreed with Einstein, arguing that our knowledge is the understanding of the very real indeterminacy in the nature of things, as well as the interdependence of subject and object. Einstein, canonized in popular culture for his theory of "relativity," was no relativist on the issue of the independent existence of the object. Philosophy and ideology, as well as physics, were involved in these discussions in the 1920s and 1930s (see chapter 7).

Chemistry and Quantum Mechanics

Among those involved in discussions at the Brussels conference was the American chemist Irving Langmuir (1881–1957), a research scientist at the General Electric research laboratory in Schenectady, New York, who was to receive the Nobel Prize in Chemistry in 1932 for his work on atomic and molecular surface phenomena. By the time of the 1927 Brussels conference Langmuir was already well known for this work as well as for his further development of G. N. Lewis's theory of the electron-pair bond. This theory had received an independent physical explanation in Pauli's principle of exclusion and in Uhlenbeck and Goudsmit's theory of paired electron spin in 1925.

Gilbert N. Lewis (1875–1946) published a theory of the electron-pair valence bond in 1916. Although little noticed at first, his paper was frequently read by the early 1920s. As in the models of Thomson, Bohr, and Walther Kossel (also in 1916), Lewis proposed that the loss or capture of electrons accounts for chemical reactivity as the atom tends to achieve the electron configuration (a ring of two or eight electrons) of an inert gas. Lewis's 1916 theory was a static model of eight electrons arranged at the corners of a cube, stressing the electron pair as the fundamental bonding unit.

This pair of electrons, Lewis proposed, is shared between two atoms in a chemical molecule. Lewis was not inclined toward the notion of internal molecular "ions" but rather insisted on gradations of electron-pair distribution between two atoms, so that the chemical bond ranges from strongly polar to nonpolar. The idea of internal molecular polarity was one that was increasingly under discussion at this time among organic chemists as well as inorganic chemists, many of them working from the postulate of J. J. Thomson, presented in his 1903 Yale University lectures, that argued that Faraday tubes of force linking atoms within the chemical molecule must arise from positive charge and end in negative charge, leading to internal molecular polarities (see figure 9).

In 1919 Langmuir gave the names *covalent* and *electrovalent* to chemical bonds, consciously extending Lewis's theory but distin-

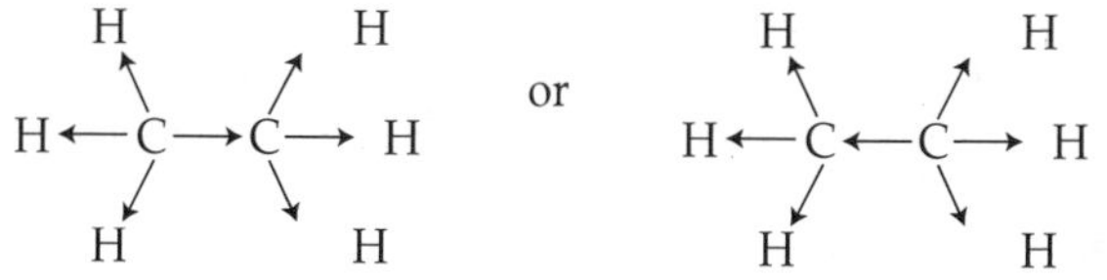

$$\mathrm{H} \leftarrow \mathrm{C} \rightarrow \mathrm{C} \rightarrow \mathrm{H} \quad \text{or} \quad \mathrm{H} \leftarrow \mathrm{C} \leftarrow \mathrm{C} \rightarrow \mathrm{H}$$

Figure 9. Directional tubes of electrical force in the hydrocarbon molecule ethane as argued by J. J. Thomson

guishing more strongly than had Lewis between two "types" of electron bonds. At a joint session of the chemistry and physics sections of the BAAS in Edinburgh in 1921, Langmuir laid out the electron-pair theory to a British and Continental audience that had mostly not heard of the American theories. "Electron rearrangement," said Langmuir, "is the fundamental cause of chemical action."[25]

Among those who was then hearing Langmuir for the first time was Arthur Lapworth (1872–1941), professor of physical chemistry at Manchester and mentor to Robert Robinson (see chapter 5). While teaching at Liverpool from 1915 to 1920, Robinson engaged in a lively and long correspondence with Lapworth about the means of applying ionic and electronic ideas in chemistry, a subject of intense interest to Lapworth from the early 1900s on.

Together they developed the idea that partial dissociation of organic molecules and the development of a slight internal molecular polarization is the crucial step in chemical reaction, just as it is for inorganic molecules in electrolytic solution. Ionization and intermediary ionized states are as important in chemistry, they argued, as in electrical and radiation physics. Lapworth gave what he called induced "alternating" polarity the primary role in initiating and determining the course of organic reaction, with the influence of a "key atom" extending over a long hydrocarbon chain if double bonds are present.

Once Lapworth told Robinson of Langmuir's Edinburgh lecture, Robinson instantly added the movement of electron pairs to the idea of internal and alternating polarity. As in his earlier correspondence with Lapworth, Robinson represented the internal dynamics of an activated molecule by arrows, implying the movement of electrons in the double bond.

The work of Lapworth and Robinson on reaction mechanisms continued for six years, and Robinson regularly taught these ideas to his advanced students at Manchester. The theory was characterized by visual imagery and useful graphic devices, such as curved arrows to indicate direction of electron displacement and noughts to denote and distinguish electrons and crosses. This notation helped chemists and their students predict the course of reaction, especially for *ortho*, *meta*, and *para* di-substitutions in aromatic rings.

By the late 1920s and 1930s Christopher Ingold was developing a rival theory and classification rooted in the Lapworth/Robinson scheme, with which he initially had voiced some substantial disagreements. Whereas Lapworth and Robinson employed a theoretical terminology rooted in traditional organic and electrochemistry (e.g., the terms *crotonoid,* or *cationoid,* and *crotenoid,* or *anionoid*), Ingold invented a simple, memorable chemical language rooted in the model of the nucleus/electron atom—for example, using the terms *electrophilic* and *nucleophilic* for predicting activities at reaction sites. Ingold also suggested a distinctive new notation [δ+, δ-] for partial or fleeting positive and negative charges that appear at the instant of internal molecular rearrangements.

Ingold, who began his university studies in chemistry in 1913 at Imperial College with the organic chemist Jocelyn Thorpe (1872–1940), was unusual among organic chemists for his interest

Sir Christopher K. Ingold back at the Chemistry Department at University College London after he was knighted by Queen Elizabeth II at Buckingham Palace in 1958. This photograph was taken by Professor Alwyn G. Davies. *Courtesy of K. U. Ingold.*

and mastery of mathematical physics and physical chemistry. He also was unusual (like Lapworth) in his focus on unifying the subdisciplines of physical chemistry and organic chemistry and, within organic chemistry, bringing together aliphatic and aromatic chemistry. Along with Louis Hammett (1894–1987) at Columbia University, Ingold is considered one of the founders of the field of physical organic chemistry. He also was one of the authors of the modern theory of chemical resonance.

In 1926 Ingold and his wife, Edith Hilda Ingold, coauthored a paper on the role of polar and nonpolar dissociation in reaction mechanisms for aromatic chemistry. One of the several important contributions of this paper had to do with the structure of the so-called conjugated molecules (molecules such as butadiene and benzene, represented with alternating single and double bonds). The Ingolds suggested that such molecules may have a ("real") structure that is different from any of the valence-bond structures that are used to represent the molecule.

In the case of benzene, for example, none of the various bond formulas (alternating single and double bonds, Kekulé formulas, Dewar formula) is truly representative of the molecule. Indeed, the Ingolds proposed, there are no true double or single bonds in benzene. (Fritz Arndt in Breslau, Howard Lucas at Caltech, and Thomas Lowry in Cambridge independently published similar ideas about conjugated molecules beginning in 1924.) A molecule like benzene is a chemical "mesomer" (between-the-forms), as Ingold expressed it. In the course of the next decade, chemical "mesomerism" was to become better known in the wider scientific community as chemical "resonance."

Here again, the physicist Heisenberg was a crucial figure in the invention of an important theory. In 1926 Heisenberg used an analogy from classical mechanics to explain the spectral lines of helium by quantum mechanics. In classical mechanics, "resonance" occurs when two systems, capable of sustaining similar oscillations, are coupled or allowed to interact, thereby perturbing each other. The coupled system will have harmonic oscillations, one with a reduced and the other with an increased frequency in comparison to the normal modes. Heisenberg applied this classical insight to the wave functions of the two electrons in a helium atom to explain the atom's stability.

In 1927 two independent studies of the hydrogen molecule appeared, both answering the century-old question of how the coexistence of two hydrogen atoms in a diatomic molecule can be consistent with an electrical theory of chemical affinity (see chapter 2).

Walter Heitler (1904–1981) and Fritz London (1900–1954) treated the H_2 molecule as two fixed hydrogen nuclei (H^+) around each of which an electron circulates. They assumed that (in zeroth approximation, or the lowest energy value) the wave function of each electron is centered on one of the nuclei—in other words, that the H_2 molecule consists of two hydrogen atoms. They showed that the two electrons, when they have antiparallel spins, tend to aggregate in the region between the two protons, thereby reducing the total energy of the system.

Heitler and London calculated how the energy of the system is dependent on the internuclear distance, and they demonstrated that the ground energy state is the singlet state and that the triplet state does not bind. In subsequent papers they specifically applied these results to chemical valence theory, deducing well-known chemical facts, including the combination of hydrogen atoms to form a hydrogen molecule, the nonexistence of a diatomic helium molecule or ion, and the fact that phosphorus, arsenic, tin, and bismuth may have valences of 1, 3, or 5 whereas nitrogen only has valences of 1 and 3.

Friedrich Hund published an entirely different approach to the binding of the hydrogen molecule, based in the mathematics of Cambridge theoretical physicist Douglas Hartree (1897–1958). Hund's work also was a generalization of the treatment of the hydrogen molecule-ion by the Danish physicist Oyvind Burrau. Rather than building up resonating wave functions from individual nucleus-centered electron waves, Hund assumed that an individual electron moves in a potential field that results from all the nuclei and the other electrons present in the molecule. As Robert Mulliken, who met Hund in Göttingen, where he was Born's assistant, later described it, this approach "regards each molecule as a self-sufficient unit and *not* as a mere composite of atoms."[26]

The first strategy came to be known as the atomic-orbital method and the second as the molecular-orbital method. By the mid-1930s Linus Pauling successfully solved an array of theoretical chemical problems using the first strategy. Meanwhile, the American Robert Mulliken (1896–1986) and the German Erich Hückel (1896–1980) refined the molecular-orbital approach, a method that favors some particular region of space and disfavors others.

In contrast to the Heitler-London-Pauling method, the molecular-orbital method overemphasizes the ionic character of a molecule. For example, for the H_2 molecule, Hund's wave-function equation assumes that it is just as probable to have two electrons around the same nucleus as to have one electron around each nucleus. For a molecule made up of identical nuclei, this treatment is a considerable exaggeration of the ionic character of the molecule.

John C. Slater (1900–1976) had studied molecular spectroscopy with Edwin Kemble at Harvard University and returned to teach there. Slater developed an approach (the "determinantal method") that offered a way of choosing among linear combinations (essentially sums and differences) of the polar and non-polar terms in the Hund-Mulliken equations. Slater's approach brought the Hund-Mulliken method into better harmony with the Heitler-London-Pauling method where polar terms do not figure in the wave equation.

Pauling's approach—articulated in his published papers and in his book *The Nature of the Chemical Bond* (1939) and lent force by his charismatic personality—earned most chemists' allegiance, at least up until 1940. Pauling had been a high-school student in Portland, Oregon, but he left without a diploma to protest having to take courses that he considered pointless. After graduating in chemical engineering from Oregon Agricultural College (now Oregon State University), he took a Ph.D. at Caltech in 1925, followed by postdoctoral studies with Sommerfeld, whom he had heard lecture in Pasadena. Pauling also spent time at Bohr's and Schrödinger's institutes.

Back at Caltech in 1927 as an assistant professor, Pauling focused on X-ray crystallography (see chapter 5). He then turned to the relation between molecular structures and the properties of compounds, attempting to model Lewis's static electron atom into a dynamic one. Fired up by his experiences in Europe, including his conversations with Heitler and London in Munich, Pauling wrote an essay review for *Chemical Reviews* on the quantum mechanics of the hydrogen molecule, explaining the work of Heitler and London and of Burrau.

At this time Pauling was already thinking about ways to mathematically explain valence and bonding in unsaturated molecules such as the nitrogen oxides and benzene. Further, he wondered how best to treat one of the simplest saturated organic molecules, methane. The carbon atom has six electrons, which, on the basis of quantum principles, should be distributed into energy states of $1s^2$, $2s^2$, $2p^2$. However, carbon has four valence electrons, suggesting that one of the 2*s* electrons has been promoted to the higher *p* energy state so that the electron configuration in carbon is $1s^2$, $2s$, $2p_x$, $2p_y$, $2p_z$.

If this were the case, one of the CH bonds (the 2*s* one) in CH_4 would be different from the others. It is not. Pauling later told historian John Heilbron that by early 1928 he thought of doing away with the distinction between 2*s* and 2*p* energy sublevels for the four shared-electron-pair bonds in methane in order to get equivalent tetrahedral valences. He mentioned the idea in a brief paragraph in a National Academy of Sciences article, but he did not yet see how to solve the problem mathematically.

By 1931 Pauling had the answer, as did Slater, who presented his own solution at conferences during the course of 1930. The basic idea was the mixing, or "hybridization," of the *s* and *p* energy levels, so that a new valence-bond wave function has a lower energy value intermediary between the energy values associated with either an *s* (spherical) or a *p* (elliptical) wave function or orbital. Pauling and Slater demonstrated that certain types of wave functions project out in characteristic directions: *p* waves, for example, can be represented by three dumbbell-shaped distributions or contour-lines at right angles to one another, and the *s* wave by a spherically shaped distribution.

Hybridization of these wave functions, or orbitals, produces electron distributions identical in kind and oriented toward the corners of a tetrahedron rather than at right angles—that is, with C-H-C angles of 109.5° rather than 90°. If a bond angle is expected to be 90° but departs from that figure (109.5° in methane, 107° in ammonia, 104.5° in water), hybridization could be suspected. The quantum mechanical calculation, Pauling said, "provides the quantum mechanical justification of the chemist's tetrahedral carbon atom"[27] (and of G. N. Lewis's simple cubes).

Pauling now turned in earnest to unsaturated molecules and to conjugated molecules with alternating single and double bonds. He extended the notion of mechanical resonance to molecules such as carbon dioxide, benzene, and graphite as well as to carbonate and nitrate ions. This approach was meant to be consistent with the chemical conceptions of Ingold's "mesomerism" and Fritz Arndt's *Zwitterion* (a "two-in-one" or "hermaphroditic" ion) but it was distinctly different from Kekulé's old hypothesis of oscillating atoms or the theory of tautomeric molecular forms in equilibrium with one another.

Pauling reasoned that a wave function might be set up to represent each of the possible classical valence bonds, or electron-pair bonds, in compounds such as carbon dioxide or benzene. Each equation corresponds to a combination of ionic and covalent character for a bond as well as to its energy content. For benzene, there are a series

Linus Pauling lecturing on the application of electron wave mechanics to the chemical valence bond at a seminar at Olso University, Norway in May 1982. *Courtesy of the Ava Helen and Linus Pauling Papers, Special Collections, Kerr Library, Oregon State University.*

of alternative representative electronic structures for the molecule, including the alternating single and double bond structures of Kekulé and the bridged structures of Dewar. The actual electronic structure is none of these, but something with elements from each.

Wave mechanics allows the calculation of the relative contribution of each bond to the structure, with the relative weights depending on chemical and physical measurements. Each of the energy values for the alternate electronic structures is higher than the energy value for the molecule, and thus the actual "resonance hybrid" is the stable form because it has the lowest energy value.

As a chemist, Pauling recognized and insisted on the origins of the resonance, or mesomerism, theory in the conceptions of classical organic chemistry:

> It is true that the idea of resonance energy was . . . provided by quantum mechanics . . . but the theory of resonance in chemistry has gone far beyond the region of application in

> which any precise quantum mechanical calculations have been made, and its great extension has been almost entirely empirical, with only the valuable and effective guidance of fundamental quantum principles.[28]

Chemists had known for a very long time that the chemical bonds in benzene are considerably more saturated than are ordinary double bonds, but also that they are very different from single bonds. Now there was reinforcing information from studies of both the chemical and the physical aspects of phenomena that current theories were on the right track, or at least on converging tracks. Chemists in the late 1920s saw the basic chemical facts and theories of nineteenth-century and early twentieth-century chemical practice confirmed in the new theoretical physics. These confirmations included the diatomic nature of the elementary hydrogen molecule, the tetrahedral directionality of carbon valence bonding, and the centrality to chemical stability of the electron duplet. X-ray crystallography similarly confirmed classical organic structure theory.

Some physicists, Dirac among them, took the view that physics is a more fundamental science than chemistry. Dirac began his 1929 paper on the quantum mechanics of a many-electron system with a statement that became somewhat infamous for its hubris: "The underlying physical laws for the mathematical theory of a large part of physics and the whole of chemistry are thus completely known."[29] Other physicists, including Dirac's mentor Ralph Fowler (1889–1944), were less arrogant. At the centenary meeting of the BAAS in 1931, in a symposium on "The Structure of Simple Molecules," Fowler cautioned his colleagues,

> One may say now that the chemical theory of valency is no longer an independent theory in a category unrelated to general physical theory, but just a part—one of the most gloriously beautiful parts—of a simple self-consistent whole, that is of non-relativistic quantum mechanics. I have at least sufficient chemical appreciation to say rather that quantum mechanics is glorified by this success than that now "there is some sense in valencies," which would be the attitude I think, of some of my friends.[30]

Pauling himself, as his interests during the 1930s turned increasingly to biologically active molecules, wrote that his views were less reductionist than they had been in his youth: "There is more to chemistry than an understanding of general principles. The chemist is also, perhaps even more, interested in the characteristics of individual substances—that is, of individual molecules."[31]

While, like Pauling, many chemists turned to the larger and larger molecules of biological significance, many physicists after 1930 concentrated on the smaller and smaller particles of the atomic nucleus in their study of relativistic quantum mechanics, nuclear physics, and solid-state physics. Some of the results, in an era of increasing national chauvinism and international political instability, are the subject of the concluding chapter.

7

Nationalism, Internationalism, and the Creation of Nuclear Science, 1914–1940

The Great War and the Chemists

At the turn of the twentieth century, Edward Frankland Armstrong was studying chemistry in Germany, just as his father, Henry Edward Armstrong, had done before him. As mentioned in chapter 1, the younger Armstrong was considerably less enthusiastic about his German experience than his father had been. Not only did Edward Armstrong find the chemical work in van't Hoff's laboratory repetitive and routine, but he also encountered strong prejudice against Englishmen. This bias, he wrote to his father, seemed to be the result of pro-Boer sentiment in Germany during the ongoing conflict between British forces and Dutch settlers (Boers) in South Africa.[1]

Scarcely a decade later, in 1908, Wilhelm II, the German emperor and the son-in-law of Queen Victoria, visited England. He startled both Germans and Englishmen by telling a *Daily Telegraph* journalist that large segments of the German population were anti-English. A steady buildup of the German navy was already fueling English diplomatic anxieties about German ambitions, just as German encroachments on French interests in Morocco were disturbing French diplomats. The assassination in Sarajevo of the Austrian Archduke Francis Ferdinand by a Serbian nationalist in June 1914 set in train the events leading to war in August, with the Central Powers of Germany, Austria-Hungary, and Turkey set against an Allied bloc that included France, Great Britain, and Russia.

French scientists had plenty of scores to settle with Germany, among them Germany's 1871 annexation of the provinces of Alsace and Lorraine just west of the Rhine. Albin Haller (1849–1925), for example, was one of the leading organic chemists and scientific administrators in Paris in August 1914. An Alsatian by birth, Haller had been a pharmacist's apprentice in Münster (in Alsace) in 1870. He had fought with the French forces during the Franco-Prussian War, and he moved to the University at Nancy in French Lorraine after 1871.

When war broke out in 1914, Haller became chief counsel on explosives to the French government, and he looked forward to a war that would restore his birthplace to French sovereignty. The Allied victory in 1918 cost him the life of his only son. It likewise cost Max Planck one of his two sons and Walther Nernst both of his sons—among the millions of lives, military and civilian, lost during the war.

Planck was rector of the University of Berlin in August 1914. He took pride in the acts and sentiments of German duty and patriotism. His friend and colleague Adolf von Harnack, a prominent historian of religion, expressed similar views: the great war would bring to the German people "one will, one force, a holy seriousness of purpose."[2] In October 1914 Planck and ninety-two other German intellectuals signed their names to a published manifesto of support for the German army and for German culture. This came in the wake of Germany's invasion of neutral Belgium, which was followed by the killing of Belgian citizens, the burning of the library of Louvain, and the bombardment of the cathedrals of Mechelen (Malines) and, in France, Reims.

Written by the playwright Ludwig Fulda, the "Appeal to the Cultured Peoples of the World" became an infamous document. It included signatures of twenty-two German scientists (among them Emil Fischer, Nernst, Wilhelm Ostwald, Conrad Röntgen, Wilhelm Wien, Richard Willstätter, and Fritz Haber) who protested "before the whole civilized world against the calumnies and lies with which our enemies are striving to besmirch Germany's undefiled cause."[3]

Among those who did not sign the document was Planck's newly appointed Berlin colleague Albert Einstein, who joined University of Berlin physiologist Georg Nicolai and two other Berlin professors in a "Manifesto to Europeans" calling for cooperation among scholars of warring nations. While Einstein's pacifist stance provided fodder for German anti-Semites (whose rantings Edward Armstrong noted in his 1900 letter to his father), those German Jewish intellectuals who, like Haber, did sign the "Appeal" derived no protection against anit-Semitism from having done so. By the mid-1930s they were equally under attack.

As the Great War began, scientists were not initially mobilized to give technical aid to the military, but they soon persuaded officials that their work could make a difference to the military effort. In France, Louis de Broglie was in charge of the army's radio transmissions from the Eiffel Tower; Jean Perrin worked on acoustical devices for the detection of mines, submarines, and airplanes; and Marie Curie directed mobile radiology units at the battlefield.

Well before the United States entered the war in 1917, George Ellery Hale (1868–1938), the director of the Mount Wilson Observatory in Pasadena and the foreign secretary of the National Academy of Sciences, offered the services of the academy to President Woodrow Wilson and formed the National Research Council at Wilson's request in 1916. Initially, there was some rivalry between this council of academic physicists and the Naval Consulting Board, headed by Thomas Alva Edison (1847–1931), which self-consciously excluded members of the American Physical Society and the National Academy of Sciences from its numbers. Asked why they were omitted, a member of the board (which included the inventors Leo Baekeland and Elmer Sperry as well as General Electric research director Willis R. Whitney) replied that the academic scientists "have not been sufficiently active to impress their existence upon Mr. Edison's mind." [4]

The Naval Consulting Board's most dramatic project was its work on the "aerial torpedo"—a flying bomb—that used a gyrostabilizer, an instrument developed by the Sperry Gyroscope Company to stabilize a ship during the firing of its guns. These aerial-torpedo studies inaugurated a new era of automation and feedback devices after the war. When launched from a distance of a hundred miles, Sperry predicted, aerial torpedoes could destroy large cities. The threat of the aerial torpedo's use would render "war so extremely hazardous and expensive that no nation will dare go into it."[5]

Less successful were the Naval Consulting Board's efforts to compete with the National Research Council in the development of acoustical methods of submarine detection. University of Chicago physicist Robert Millikan (1868–1953) persuaded the navy to bring some ten university physicists to New London, Connecticut, where they developed detection devices that were placed on submarine-chasers in the English Channel and the Adriatic. By the time United States forces had landed in Europe, American physicists had also figured out procedures for sound-ranging and flash-ranging that could be used for precise detection of artillery sites in the battlefield.

But physicists' laboratory work turned out to be neither as dramatic nor as problematic as that performed by chemists. During the earliest stage of the war, chemists, too, found that their services were not much sought-after by the military. Victor Grignard (1871–1935), for example, a Nobel Prize winner in chemistry in 1912, was near his family home at Cherbourg, France when the war broke out, and so he joined the territorial regiment. After an officer noticed the red ribbon of the Legion of Honor on Grignard's uniform and some of his Parisian colleagues intervened out of fear he would wind up in the trenches, Grignard was assigned to a Paris office for chemical war matériel.

Gradually, the Great War became known as "the chemists' war." One of the chemists' principal jobs was the production of explosives, including "disruptive" shells—high-explosive shells that detonated on impact. British and French chemists relied largely on picric-acid explosives, while the Germans used trinitrotoluene (TNT), which was readily available because of German superiority in the coal-tar chemicals industry. One of Grignard's first studies during the war in his own laboratory at Nancy focused on cracking heavy fractions of benzene to guarantee a French source of toluene.

In both Allied and Central Power countries, as well as in the United States, scientists were aware before their governments of the role scientific research and applications might play in time of war. Fritz Haber (1868–1934), for example, worried well before August 1914 that Germany's raw materials would be cut off in time of crisis, warning that, "it is coal for which countries will wage war."[6] Haber's efforts to forge a connection between the Prussian War Ministry and his own Kaiser Wilhelm Institute for Physical Chemistry, (opened in 1912) initially came to nought, but he received more positive responses about funding after the war began.

The terrible and unforeseen stalemate of military action into a state of trench warfare, with troops pinned down by machine-gun fire, helps to account for the resort to gas warfare. In December 1914 reports came in that Germans were throwing gas grenades into the trenches at the Lorraine front. The first gas used was bromoacetic ester, a tear gas. The gas next employed by the German army—at Ypres

Victor Grignard, Nobel Prize 1912 and Chevalier of the Legion of Honor, on guard duty in Cherbourg at the beginning of World War I. From Roger Grignard, *Centenaire de la naissance de Victor Grignard 1871–1971* (Lyon: Audin, 1972). *Courtesy of Mme. Colette Grignard.*

in April 1915—was deadly chlorine gas, produced in Fritz Haber's Dahlem laboratory under the chilling code-name "Disinfection." Chemists on both sides worked to develop adsorbent and protective masks against the gases, as well as to study the properties of toxic gases. Grignard's group prepared phosgene ($COCl_2$) by the action of oleum on carbon tetrachloride, and the French army was soon employing phosgene against the Germans.

Research by Haber's group resulted in a different kind of gas weapon, used at Ypres on 12 July 1917. This one caused horrible blisters and incapacitated a soldier simply by coming into contact with his skin and lungs. The gas disabled as well as killed its victims. Within four days Grignard and other French chemists in Paris realized that the new gas was dichlorethyl sulfide, or mustard gas, which had first been prepared by the English chemist Frederick Guthrie in 1860. By the winter of 1918, after a trip to the United States, Grignard had developed a precise test for mustard gas that was first employed at the battlefield just before the Armistice. By that time there had been approximately one million gas casualties.

The horrors of World War I left scientists and engineers with a foreboding about any war that might follow, a war that would use not only highly explosive artillery but toxic gases, airplanes, and submarines. As one French scientist put it, "The war of tomorrow will surpass the horror that has preceded it; without any doubt, the perfection of aviation and chemical warfare will allow entire regions to be rendered uninhabitable."[7]

As a consequence of these fears, the Geneva Convention, adopted by international delegates in 1925, outlawed the use of gases in warfare. The United States, with strong advice from many American chemists, declined to sign the Geneva treaty on the grounds that former enemy nations could not be trusted to abide by its terms and would produce the banned gases secretly. Some American chemists also agreed with the argument of General William L. Sibert, director of the wartime Chemical Warfare Service, that far fewer soldiers and civilians died from gas than from rifles and artillery. The United States ratified the Geneva treaty only in 1974.

While much has been written about atomic scientists' sense of responsibility for the death and destruction in Hiroshima and Nagasaki, many chemists brooded over their responsibility in the earlier Great War. Fischer wrote reflectively in a personal letter that although, unlike Haber and Nernst, he had declined to play a role in the development of poison gases, he nonetheless worried that he had indirectly contributed to the terrible war through his central position in organic chemistry, which had in effect created the coal-tar and high-explosives industries. And Haber, who in 1917 had directed a staff of fifteen hundred engaged in work on chemical war gases, lived ever afterward with the personal sorrow of his first wife's suicide. In what many viewed as a self-conscious protest against Haber's promotion of poison gas as a weapon of war, she had used his service revolver to kill herself.

The Failure of Internationalism in the 1920s

With the war over and the Treaty of Versailles signed, the Swedish Academy of Sciences announced the winners of the Nobel Prize in

Chemistry for 1918 (none was designated for 1919) and the Nobel Prizes in Physics for 1918 and 1919. To the astonishment of scientists in the former Allied countries, the recipients were Haber, Planck, and Johannes Stark. Not only were they all Germans, but Haber was widely vilified outside Germany as the author of gas warfare. The synthesis of ammonia from its elements, for which Haber was awarded the prize, was an industrial process that had helped Germany produce its own nitrates for explosives during the war. French scientists for the first time declined to attend the annual ceremony in Sweden.

International cooperation among scientists did not soon revive after the war. The League of Nations' International Commission on Intellectual Cooperation (Commission internationale de cooperation intellectuelle, or CICI), which first met in 1922 in Geneva, excluded German and Austrian scientists. When the British physical chemist F. G. Donnan and the Dutch physical chemist Ernst Cohen convened a summer meeting in 1922 to discuss reestablishing relations among chemists of the former Allied and Central blocs, French chemists declined to attend.

Before the war, the Belgian industrialist Ernest Solvay had wanted to establish international physics and chemistry institutes in Brussels. In autumn 1911, as noted in earlier chapters, the first Solvay Congress of Physics convened in the Hôtel Métropole with some two dozen Continental and British physicists and physical chemists in attendance. In 1911 and 1912, Ernest Solvay corresponded with William Ramsay and Wilhelm Ostwald about setting up a similar chemistry conference, but nothing happened until after the war.

By then, when some twenty-five chemists convened in Brussels in 1922, no German or Austrian scientist was welcome. Not until the fourth Solvay Chemistry Congress in 1931 was a chemist invited from any German-speaking country other than neutralist Switzerland. In the postwar Solvay Physics councils, Schrödinger was the only physicist from the former Central Powers who attended a meeting before 1927.

Einstein was invited to meetings in 1921 and 1924 but did not attend, in the first instance because he was visiting the United States at the time of the meeting and, in the second instance, because he had decided not to participate as long as his German colleagues were excluded. "In my opinion it is not right to bring politics into scientific matters, nor should individuals be held responsible for the government of the country to which they happened to belong," he explained.[8]

Immediately after the war George Ellery Hale, the British physicist Arthur Schuster (1851–1934), and the French mathematician Charles-Émile Picard (1856–1941) were among the scientists involved in reviving the prewar International Association of Academies as a new International Research Council (CIR). Against the wishes of U.S. president Woodrow Wilson, with whom Hale discussed these matters, the council's statutes excluded German and Austrian academies from membership. Picard argued, unsuccessfully, for keeping out neutralist countries' academies, including those of Sweden, Denmark, and the Netherlands, on the grounds that they would only serve as an entering wedge for the Germans. While political accords were signed between German and French educational ministers in 1925 to foster scientific exchange, another year passed before Germany joined the League of Nations and German academies were invited to join the CIR.

Once invited to join, the impasse was not over. Germans never joined the International Research Council because their conditions for guaranteed positions of leadership were not met. Nor did the Germans join the CICI when invited in 1924. The Verband der Deutschen Hochschulens (VDH), an organization of German university-level institutions, established rules forbidding German scholars' communications with foreign friends unless their correspondents renounced the International Research Council's position. The Notgemeinschaft der Deutschen Wissenschaft, an organization of German academies and institutes that included the VDH, formally rejected any statement of German responsibility for the war and insisted its members act to safeguard "German dignity."[9] Internationalism was a real casualty of the war.

Chemists and Physicists in a Postwar Political Maelstrom

The only German chemist invited to any of the Solvay Chemistry councils in Brussels during the 1920s (in 1925) was Hermann Staudinger, who had moved from the University of Strassburg to the Zurich Polytechnic Institute (ETH) in 1912. During his twelve years in Zurich, Staudinger and his coworkers studied ketenes, pyrethrines, and the polymerization of isoprene. Around 1920, Staudinger began to develop the controversial theory of macromolecules based on studies of polyoxymethylene, natural rubber, and polystyrene (see chapter 5). Among his coworkers were two future Nobel Prize winners, Leopold Ružička (1939) and Tadeus Reichstein (1950).

During the war Staudinger synthesized a pepper substitute and isolated the compounds that give coffee its aroma. He and his first wife Dorothea Förster became members of a pacifist circle organized by Leonhard Ragaz, an evangelical Zurich socialist, and they corresponded with Georg Nicolai, who with Einstein had signed the "Manifesto to Europeans." Shortly after the United States entered the war in 1917, Staudinger wrote a personal letter to German general Erich Ludendorff calling for peace negotiations. He also encouraged the International Committee of the Red Cross in Geneva to make a formal appeal to the Allies and Central Powers for an end to gas warfare. Staudinger's denunciation of chemical warfare appeared in the *Revue internationale de la Croix-Rouge* in May 1919 just before Germany signed the Versailles treaty.

Haber was among many German colleagues who accused Staudinger of betraying "Germany at the time of her greatest crisis and helplessness." Haber denied that the use of poison gas had violated international law and claimed that French troops had used it first. When the University of Freiburg offered Staudinger a professorship in chemistry, Staudinger found himself required by Freiberg adminstrators to explain his "political past."[10]

Staudinger's past activities again became suspect in 1933, after Adolf Hitler's appointment as chancellor. Philosopher Martin Heidegger, rector of the University of Freiberg, advised the Gestapo to investigate Staudinger, who began to consider, but decided against, emigrating to the United States. Among those chemists who continued to oppose his chemical theory of macromolecules in the 1930s and who successfully blocked some of his research funding and publications were Kurt Hess, a member of the Nazi Storm Troopers (SA), and Wolfgang Ostwald, an adviser to the Nazi government on academic affairs. Staudinger nonetheless continued his laboratory work and in 1939 became editor of the *Journal für praktische Chemie* (Hermann Kopp's old journal, which eventually became the journal *Makromolekuläre Chemie*, published in Basel).

Others fared considerably worse than Staudinger. In August 1920 the anti-Semitic National Socialist agitator Paul Weyland delivered a public lecture in Berlin slandering Einstein, a Jew, as a pacifist and accusing him of plagiarism, publicity-seeking, and scientific "dadaism." After Einstein accepted an invitation to give a plenary lecture at the centennial celebration of the Society of German Scientists and Physicians in 1922, he received not only death threats but the appalling news of the murder, on 24 June 1922, of his friend

Walther Rathenau, foreign minister of the Weimar Republic. Einstein withdrew from the lecture.

Only a few months earlier, he had been urged by Rathenau to accept, in the interests of internationalism and peace, an invitation from Paul Langevin to lecture in Paris. So controversial was the appearance of a notable German-Jewish scientist in Paris that French police had to cordon off the area around the Collège de France in order to prevent interference from the right-wing anti-Semitic organization Action Française. Only those who arrived with personal invitations were admitted to Einstein's lecture hall.

While some German ministers in the early 1920s described Einstein as "just now a cultural factor of the first importance for Germany,"[11] other Germans expressed rage that he was to receive the Nobel Prize in Physics for 1921. The German physicist Philipp Lenard, a Nobel Prize winner in 1905 for his experimental investigations of cathode-rays, protested to the Swedish Academy's physics prize committee that Swedish scientists "had not been able to bring to bear a sufficiently clear Germanic spirit to avoid perpetrating such a fraud."[12]

After being awarded the Nobel Prize in Physics for 1919, Stark relentlessly carried out a personal vendetta against Einstein. In 1922 he published a pamphlet entitled *The Contemporary Crisis in German Physics,* which was punctuated with vilifying characterizations of "Jewish science" and "Jewish theoretical physics"—which Stark claimed were undermining the character of true German experimental science. Relativity theory was Stark's prime example of this alleged degradation.

While Planck and many other Germans said that Hitler would have to moderate his earlier political positions once he assumed the chancellorship in January 1933, the civil service laws of 7 April 1933 demonstrated that the chancellor and his party fully intended to "cleanse" the nation of non-Aryan Jews and Gypsies and others perceived as threats to Germany, including Social Democrats and Marxists. The initial civil service laws of spring 1933 required the dismissal of any public official of non-Aryan descent, meaning anyone with a Jewish parent or grandparent. A prison camp was set up in 1933 just outside Munich at Dachau for forty-three leftist opponents of Nazi policies.

A public statement by Einstein in Pasadena, California, in early March 1933 had caused great furor in Germany. Einstein said that he would not return to Germany because the country's citizens no longer enjoyed "civil liberty, tolerance, and equality."[13] This statement was judged by the secretary of the Berlin Academy of Sciences to constitute slander against the fatherland. Einstein immediately resigned both his membership in the academy and his German citizenship (for the second time).

Anti-Semitism was by no means unique to Germany, although official anti-Semitism had been part of the legal code in most German states until civil and legal equality was guaranteed Jews with the establishment of the North German Confederation in 1867. In 1871 civil service positions in all of the German empire, including faculty positions in German universities, became open to Jewish citizens in principle.

The 1933 anti-Semitic laws sought to undo the egalitarian legal treatment accorded Jews for the previous sixty-two years. In March 1933, the Third Reich's new laws immediately required the resignation of many civil employees, including seven members of the sciences faculty at the University of Göttingen. The mathematician Richard Courant (1888–1972), who sought exception on the grounds that he had been a frontline soldier at Verdun in the Great War, submitted a petition signed by twenty-two German professors, including Planck and Laue. But Courant's case found no sympathy from the Ministry of Education.

James Franck (1882–1964), a Nobel Prize winner, simply resigned from the University of Göttingen and moved to a position at the Johns Hopkins University in Baltimore. Sadly, he resigned from Hopkins and left for the University of Chicago because of Hopkins president Isaiah Bowman's prejudice against Jewish faculty. David Hilbert (1862–1943), seated at a banquet in 1934 with the new minister of education, Bernhard Rust, was asked whether it was true "that your Institute suffered so much from the departure of the Jews and their friends?" Hilbert replied, it is said, "No, it didn't suffer, Herr Minister. It just doesn't exist any more."[14]

By 1936 more than sixteen hundred scholars, about one-third of them scientists, had been forced to resign their positions or had chosen to leave because of circumstances in Germany. Scientific academies in France, the United States, Great Britain, and other countries organized lists of refugees and attempted to find them temporary or permanent jobs. Many Jewish Germans found themselves moving from country to country as temporary positions expired or, after 1939, as German troops began advancing into France, the Netherlands, and Denmark.

Following the annexation of Czechoslovakia and Austria, Central European Jews who were not German citizens became subject to German laws. Among them was Lise Meitner (1878–1968) who in 1938 left the Kaiser Wilhelm Gesellschaft for Sweden. In that same year Enrico Fermi (1901–1954) received permission from Benito Mussolini's fascist government in Italy, now an ally of Germany, to travel to Sweden to receive the Nobel Prize in Physics. Fermi continued onwards, with his Jewish wife, to the United States. By this time, emigrés to the United States included Leo Szilard (1898–1964) at Columbia University, Eugene Wigner (b. 1902) at Princeton, Edward Teller (b. 1908) at Georgetown, and Victor Weisskopf (b. 1908) at Rochester, among many others. Haber left Germany for England in 1933 and died the following year in Basel.

These same decades—the 1920s and 1930s—were ones in which the "new physics" of quantum theory and relativity theory was being enthusiastically announced by journalists and embraced by some philosophers and scientists. The new physics was said to affirm a new philosophy of freedom from the old "dogmatisms" of materialism and determinism. Quantum physicists, most notably those associated with Arnold Sommerfeld at Munich, Max Born at Göttingen, Erwin Schrödinger at Zurich, and Niels Bohr at Copenhagen, self-consciously thought of themselves as making a revolution in physical laws that had direct implications for transforming notions of causality and determinism. For them, this was a "golden age" of science.

But many of the same scientists who were devoting themselves with such enthusiasm and intelligence to these exciting new theories were also worried, frightened, and oppressed—by monetary inflation and economic collapse; by Bolshevist, trade union, and socialist movements; by the rise of fascism in Italy and Nazism in Germany; and by the explosion of anti-Semitism and pan-German nationalism. British and American political movements for women's suffrage, as well as the increasing agitation for greater educational and professional opportunities for women, only added fuel to a volatile political climate, as did debates over policies of disarmament and pacifism in response to the horrors of the Great War.

While German scientists tended to avoid political activities and American scientists found themselves, for all their travels abroad, living in a national political climate of isolationism, influential French and British scientists entered into the political fray in the 1920s and 1930s. In France, for example, Perrin, Langevin, and

Frédéric Joliot (1900–1958) were among French intellectuals marching in a united Popular Front of socialists and communists in favor of peace after the 1935 Franco-Soviet pact.

Hitler's occupation of the Rhineland and the outbreak of civil war in Spain in 1936, however, split opinion among those who had been arguing for peace over the previous fifteen years. Many French and British pacifists reluctantly began preparing themselves emotionally for the outbreak of war and advising their governments of the need to enlist scientific and engineering personnel for the study of offensive and defensive armaments.

Einstein was one of the first to break ranks with his pacifist, internationalist scientific colleagues. This he did as early as July 1933, when, in response to an appeal for him to speak out on behalf of two conscientious objectors in Belgium, he said, "What I shall tell you will greatly surprise you . . . Imagine Belgium occupied by present-day Germany. Things would be far worse than in 1914 . . . Hence I must tell you candidly: Were I a Belgian, I would not, in the present circumstances, refuse military service."[15]

Einstein and Modern Physics

Among members of an elite group in international physics, Einstein was respected and admired by the time he was in his late twenties, following a series of brilliant papers he published in the *Annalen der Physik* in 1905. At the age of thirty-one he achieved the title of professor at the German University of Prague, and at thirty-five he held an unusually prestigious position tailored precisely for him in Berlin.

For the broader public, the newspapers of November 1919 were what brought Einstein's name into enduring and common parlance and conversation. In an article headlined "The Fabric of the Universe," the 7 November 1919 issue of the London *Times* reported on a joint meeting of the Royal Society of London and the Royal Astronomical Society during which the results of a recent British eclipse expedition were announced.

The Royal Society's president, J. J. Thomson, heralded the British expedition's confirmation of Einstein's theory of relativity as "a whole continent of new scientific ideas." The newspaper article concluded, "It is confidently believed by the greatest experts that enough has been done to overthrow the certainty of ages, and to require a new philosophy that will sweep away nearly all that has been hitherto accepted as the axiomatic basis of physical thought."[16]

Reporters knew little in the autumn of 1919 about the great discoverer of the "new continent" except that he had not signed the infamous 1914 manifesto, that he was either Swiss or German, and that he was an ardent Zionist. A followup article in the *Times* made clear that Einstein had just changed the world:

> The ideals of Aristotle and Euclid and Newton which are the basis of all our present conceptions prove in fact not to correspond with what can be observed in the fabric of the universe. . . . Space is merely a relation between two sets of data, and an infinite number of times may coexist. Here and there, past and present, are relative, not absolute, and change according to the ordinates and coordinates selected.[17]

Gradually, more came to be known about Einstein. Born in Ulm, Germany, and educated during his childhood in Munich, Einstein was raised with his sister in a secular Jewish family. His father and uncle were engineers. Hostile to the authoritarian and military style of German public education and social values at the end of the century, he was glad to leave Germany, first for Italy and then for Switzerland.

In 1903 he married Mileva Marić a fellow student at the Zurich Polytechnic Institute. She had grown up in Vojvodina, which then was part of southern Hungary. Her plans to complete a degree and become a teacher never succeeded, although the young couple's first years together were spent in enjoyment of their mutual scientific interests and aspirations.

In 1907 Einstein became a *Privatdozent*, or instructor, at the University of Bern. After his friend Friedrich Adler withdrew from competition in Einstein's favor, he filled the post of "extraordinary" (assistant) professor at the University of Zurich. There he taught for two years, before moving to the German University of Prague in 1911, and then returning to Zurich and to his *alma mater*, the Zurich Polytechnic Institute in 1912.

Courted by Planck, Nernst, and Max von Laue (1879–1960), Einstein in 1913 accepted a magnificent offer of election to the Prussian Academy of Sciences, a research position at the Kaiser Wilhelm Institute for Physics, and a professorship at the University of Berlin, with teaching duties when he wanted. He reinstituted his German citizenship in 1919.

Mileva Einstein, however, could not accommodate herself to life in Berlin and left for Switzerland with their two sons in the summer of 1914. She never returned. In February 1919 they divorced and Einstein married his widowed cousin Elsa. When Einstein received the Nobel Prize in 1922, he transferred the prize money to Mileva, as had been promised in the divorce agreement. Their daughter Lieserl, whose birth in 1902 became publicly known only in 1987 following publication of previously unpublished letters, was either adopted before their marriage or died in infancy.

By 1919 Einstein's achievements in theoretical physics were extraordinary. His "special relativity" paper was one of four that he published in the *Annalen der Physik* in 1905. One of the others, "A New Estimation of Molecular Dimensions" served as his Ph.D. thesis at the University of Zurich. Another, "On the Motion, Required by the Molecular Kinetic Theory of Heat, of Small Particles Suspended in a Stationary Liquid," became known as his paper on Brownian motion, establishing the validity of the kinetic theory of gases (see chapter 4). The first to appear, "On a Heuristic Viewpoint Concerning the Production and Transformation of Light," provided a new way of thinking and a new confirmation for the quantum theory (see also chapter 4).

Einstein's fourth paper on what later came to be called "special relativity," was entitled "On the Electrodynamics of Moving Bodies." At the time it was completed, Einstein was working as a technical expert in the Swiss patent office in Bern, where he was hired because of his knowledge of electricity, an expertise he had obtained from his studies at the Zurich Polytechnic Institute, from which he had graduated in 1900.

Einstein's work was by no means the only investigation at this time of the electrodynamics of moving bodies, nor was he alone in thinking that resolving problems in electrodynamic theory might require fundamental changes in the principles of physics.

Henri Poincaré, one of the most respected mathematicians of the early twentieth century, had been the first French scientist to introduce J. C. Maxwell's equations into French lecture halls. By 1898 Poincaré was questioning classical Newtonian assumptions about universal space and time. In 1904, at the international scientific congress in St. Louis, he delivered a widely discussed speech in which he laid out paradoxes in the kinetic theory of gases ("reversiblity in the premises and irreversibility in the conclusions") and experimental failures to measure a body's motion through the ether. To resolve the incongruity between these negative results and the requirements of Maxwell's equations, Poincaré said, "A whole new mechanics might need to be constructed in which inertia increases with velocity, the velocity of light becomes an impassable limit, and ordinary mechanics is only a first approximation for velocities not too great."[18]

Poincaré published papers in 1905 and 1906 on the dynamics of the electron, noting the experimental impossibility of demonstrating the absolute motion of the earth. Using Henrik Lorentz's paper of 1904 as a starting point (see chapter 3), Poincaré gave the name *Lorentz transformations* to Lorentz's fundamental equations relating the coordinate positions of an electron to its velocity and the velocity of light.

Poincaré was no revolutionary, however. He saw no need to call these steps anything more than necessary adjustments or complementary hypotheses to Newtonian mechanics, which must be made in recognition of experimental failures to establish absolute motion or simultaneity. Indeed, much to Einstein's later frustration, Poincaré never accepted the more radical premises of Einstein's 1905 paper on the electrodynamics of moving bodies.

In that paper, Einstein dramatically departed from his contemporaries' approach to the same problem. The paper began with a statement of an asymmetry, or paradox, within the widely held theory of electromagnetism. This asymmetry lay, he argued, in describing states of rest or motion by saying that change in an electric field is always accompanied by a magnetic field, and that change in a magnetic field is always accompanied by an electric field, as if one were at rest and the other were in motion by reference to an absolute coordinate system.

In Newtonian mechanics, Einstein argued, no experiment can distinguish whether a body is at rest or in uniform motion. Only accelerated motion can cause effects that are detected by the observer. Using light to detect the uniform motion of a body through an absolute reference frame such as the ether is just a means of cheating on the fundamental laws of mechanics. What Lorentz and Poincaré had taken to be the fact of the failure of detecting uniform motion must be transformed into a principle of the impossibility of detecting uniform motion.

The laws of electrodynamics, Einstein asserted, like the laws of Newtonian dynamics, are the same in all frames of reference. This means that, as a premise, an experiment can no more detect absolute rest or uniform motion for electromagnetic phenomena than for inertial masses. And, as another premise (not an experimental result to be explained away), light is always propagated in empty space with a velocity independent of the state of motion of the body emitting the light.

Experimental predictions followed. No material object can travel as fast as light. There is no universal simultaneity of events. The properties of a body depend on its velocity. As a mass—for example, an electron—approaches the speed of light, its length decreases, its mass increases, and the time it takes to travel through a given distance is greater than expected by the observer "at rest." Einstein calculated precise transformations from the equations of Maxwell, without realizing that these had already been published by Lorentz the previous year and had been communicated by Poincaré to the Paris Academy a few weeks earlier.

An increase in the mass of an electron traveling at high speed had been found in 1902 by Walter Kaufmann (1871–1947) at Göttingen and by others who studied deflections of beta rays emitted by radioactive sources in electrical and magnetic fields. Kaufmann, however, came to think that his data did not support Einstein's theory, and he argued instead that the data favored Max Abraham's theory of an electron's constant volume. By 1916 additional experiments by other investigators finally gave confirming results for what came to be known as the Lorentz-Fitzgerald transformations.

In 1906 Einstein published a statement that the inertia of a body should be understood to depend on its energy content. "The law of conservation of mass is a special case of the law of conservation of energy," he wrote.[19] A body traveling at a velocity near the speed of light can only take in additional kinetic energy by an increase in its mass, he argued. At velocities near the speed of light (c), the change in mass is equal to the amount of kinetic energy divided by c^2. Thus, he concluded, mass and energy are interchangeable in a precisely calculable way. In an article the following year (1907), Einstein wrote the equation

$$E = mc^2$$

for this relationship, with the presumption that radioactive and nuclear phenomena might eventually provide a test for its validity.

Einstein was discontent with his 1905 electrodynamic theory, however, because it was limited to the case of bodies in states of rest or uniform motion. What about bodies in accelerated motion? Isn't there an asymmetry in the distinction between bodies accelerating free of gravitational force and bodies accelerating under the influence of gravitational force? How, Einstein asked, could Newton's classical concept of action at a distance coexist with Maxwell's concept of the structure of the field?

> In Maxwell's theory there are no material actors. The mathematical equations of this theory express the laws governing the electromagnetic field. They do not, as in Newton's law, connect two widely separated events; they do not connect the happenings *here* with the conditions *there*.[20]

Relativity and the Quandary of Causality

From 1907 to 1911 Einstein found himself working much of the time on the problem of acceleration in an attempt to develop a more general theory in which no distinction is made between kinds of motion as in Newtonian mechanics. He found himself thinking about Ernst Mach's criticism of the Newtonian coincidence that inertial mass and gravitational mass are identical, a coincidence that neatly explains Galileo's strikingly counterintuitive argument that bodies of different densities falling from a tower in the absence of resistance travel at the same speed.

For Newton, since inertial force is defined by mass and acceleration,

$$F_{inertial} = m_1 a$$

and gravitational force is described in terms of attracting masses at a distance *r* from one another,

$$F_{gravitational} = \frac{m_1 m_2}{r^2}$$

Then, for a body accelerating toward the earth, where the inertial force is the gravitational force,

$$m_1 a = \frac{m_1 m_2}{r^2}$$

and the acceleration of the body m_1, as it falls freely to the earth m_2, is independent of its mass.

In 1907 Einstein turned this puzzle of equivalence of the inertial and gravitational mass into a fundamental principle for a new unified theory of electrodynamics and gravitation. Later, in an unpublished manuscript originally written as a long article for *Nature,* Einstein recalled "the happiest thought of my life":

> The gravitational field has only a relative existence in a way similar to the electric field generated by magnetoelectric induction. *Because for an observer falling freely from the roof of a house there exists*—at least in his immediate surroundings—*no gravitational field.* Indeed, if the observer drops some bodies then these remain relative to him in a state of rest or of uniform motion. . . . The observer therefore has the right to interpret his state as "at rest." . . . The experimentally known matter independence of the acceleration of fall is therefore a powerful argument for the fact that the relativity postulate has to be extended to coordinate systems which, relative to each other, are in nonuniform motion. [Emphasis original.][21]

In 1907 Einstein recognized, then, that as far as mechanical effects are concerned, an observer of accelerated motion cannot distinguish between gravitational and inertial effects—for example, whether a body is accelerating in a motionless reference frame under the influence of a gravitational "force" or whether it is accelerating in reaction against an accelerating motion of the moving reference frame.

Further, in contrast to measurements taken for a coordinate system in uniform motion, the velocity of light in an accelerating or gravitational coordinate system is found not to be constant as measured by observers with measuring sticks and clocks. In a strong gravitational field, oscillations or frequencies of vibration of a mass will appear to be slower than in gravity-free space: the wavelengths of light reaching us from the sun will be shifted toward the red end of the spectrum in comparison to the wavelengths of similar atoms observed on earth.

And, as Einstein realized clearly by 1911, starlight passing near the sun should appear to an observer to be bent downwards toward the sun, just as it would appear to deviate from a straight path into a curved path if passing through an accelerating reference system. Einstein calculated the deflection, one that might be detected during a solar eclipse, at 0.83 seconds of arc (soon recalculated at 0.87 seconds).

By 1916 Einstein's general theory of relativity had gone through detailed rethinking and refinement, largely through the help of his Swiss friend Marcel Grossman (1878–1936), with whom he jointly published papers in 1913 and 1914 before Einstein left Zurich for Berlin. Others, too, were working at this time to develop a gravitational theory, among them Gunnar Nordström, Max Abraham, and Gustav Mie.

Grossman, who was professor of mathematics at the Zurich Polytechnic Institute, taught Einstein how to apply to his physical ideas the mathematics of the tensor calculus, which allowed him to treat all coordinate systems equally without playing favorites, using a principle of general covariance for all transformations, not just linear transformations.

The result was the identification of gravitation with an intrinsic curvature of a linked space-time, a geometry of points in space developed and discussed in 1908 by Göttingen professor Hermann Minkowski (1864–1909). In this geometry, a point has duration in time, represented as a world line, and curvature of time results from the varying speed of light in a gravitational field. For Einstein the metric tensor becomes the carrier of gravitation. Planets move in orbits around the sun not because the sun attracts them but because in the curved space-time around the sun there are no straight world lines. Gravitational and inertial motion are one and the same thing.

The 1916 version of Einstein's theory resulted in a prediction of an additional advance along the orbital ellipse of the planet Mercury's perihelion (the perihelion is the closest point in a planet's orbit to the sun). Einstein's computed value was approximately 43 seconds of arc per century. This was remarkable since a discrepancy of about 40 to 50 seconds within the observed perihelion advance (approximately 5,600 seconds total) was recognized as an anomaly that could not be explained on the basis of Newtonian mechanics. The fact that Einstein's theory predicted the anomaly so precisely and in the right direction was announced with considerable acclaim by the Prussian Academy of Sciences.

Einstein wrote to Paul Ehrenfest in January 1916, "Imagine my joy at the feasibility of the general covariance and at the result that the equations yield the correct perihelion motion of Mercury. I was beside myself with ecstasy for days."[22] The revised theory also gave a new figure for the deflection of the bending of starlight near the sun, approximately 1.7 seconds of arc, a figure soon refined to 1.74 seconds.

Following Einstein's 1911 derivation of the deflection effect, he had looked forward to a test of the theory during a solar eclipse. In the summer of 1914, Erwin Freundlich (1885–1964) of the Berlin Observatory mounted an expedition to southern Russia, where astronomers from Argentina and the Lick Observatory also planned to study a total solar eclipse on 21 August. Unfortunately, with the outbreak of the war, Freundlich found himself interned and his instruments impounded, although he was lucky enough to be exchanged by the Russians for Russian prisoners in September. No eclipse observations were made by anyone in Russia.

The next total solar eclipse for which photographs were taken that could be used to test the deflection prediction occurred at Goldendale in the state of Washington on 8 June 1918. But William Campbell (1862–1938) of the Lick Observatory reported to the Royal Astronomical Society in July 1919 that the results ruled out the value predicted by Einstein's theory.

By this time Arthur S. Eddington (1882–1944), secretary of the Royal Astronomical Society, had received from Willem de Sitter (1872–1934) a copy of Einstein's paper on gravitational field equations and had arranged for a series of papers by de Sitter to appear in the *Monthly Notices* of the society. Eddington himself derived the 0.87 deflection value by assuming that light of energy E has mass E/c^2. A Newtonian interpretation for this value could be given, but this interpretation did not predict the extra advance in the perihelion of Mercury, a phenomenon rederived independently by Karl Schwarzschild and J. Droste (the "Schwarzschild solution") using Einstein's field equations.

A solar eclipse on 29 May 1919 provided a new opportunity for confirmation, and two British expeditions set off in the spring, one group bound for Sobral, about fifty miles inland from the coast of Brazil, and the other—Eddington along—for Principe, an island off the western coast of Africa. The Sobral group carried an astrographic telescope and a refracting telescope with a 4-inch aperture. The Principe group had an astrographic telescope. Philosophers John Earman and Clark Glymour, who have reanalyzed the data, claim that it did not clearly support or refute the theory: the Sobral astrographic plates gave values of either 0.97 or 1.40 seconds depending on the scale-values used from the original plates or the check plates. The 4-inch telescope used at Sobral produced photographs that were better in terms of image quality and dispersion of measurement, giving 1.98 seconds, significantly above Einstein's value. The Principe result was 1.61 seconds, and it was this figure that Eddington subsequently tended to emphasize.

Eddington, a Quaker and a pacifist, was proud that this British work prevailed in the face of the strong anti-German sentiment felt by many of his scientific colleagues. In a later tribute to Frank Dyson (1868–1939), the Astronomer Royal who had initially suggested the 1919 expeditions, Eddington wrote,

> The announcement of the results aroused immense public interest, and the theory of relativity which had been for some years the preserve of a few specialists suddenly leapt into fame. Moreover, it was not without international significance, for it opportunely put an end to wild talk of boycotting German science. By standing foremost in testing, and ultimately verifying, the "enemy" theory, our national Observatory kept alive the finest traditions of science; and the lesson is perhaps still needed in the world today.[23]

Whether or not the results were clear-cut, Einstein's theory was triumphant. Other eclipse expeditions after 1919 obtained a variety of results, most of them a little higher than the value of 1.74 seconds. Red-shift data, which had not previously been judged to give a clear decision, gradually came to be seen in the 1920s as confirming the red shift of spectral lines of iron, titanium, and cyanogen in the sun.

For the rest of his life, Einstein sought to develop a unifying and secure framework for describing matter, light, electromagnetism, and gravity in a "unified field theory."

The youthful Einstein, as he wrote retrospectively in his autobiographical notes, had fallen under the sway of the phenomenalist and sensationalist epistemology of Mach. But, partly as a result of grappling with relativity theory, Einstein began to see a weakness in Mach's apparent belief that science is principally an ordering of empirical data, including our sensations. The physicist and historian Gerald Holton writes that Einstein adopted a position of "rationalistic realism." As Einstein expressed his views in his autobiographical notes, "Physics is an attempt conceptually to grasp reality as it is thought independently of its being observed. In this sense one speaks of 'physical reality.'"[24]

Einstein's views gradually became more and more at odds with contemporary developments in quantum mechanics and with the interpretations of causality and uncertainty defended by Bohr, Heisenberg, and Pauli. In Einstein's opinion, the new quantum mechanics made the world into nothing more than a cosmic craps game. But, he declared, "God does not play dice."[25] Indeterminacy, in Einstein's view, was the result of the present incompleteness of physical theory, a point of view he argued in a paper coauthored with Boris Podolsky and Nathan Rosen at the Institute for Advanced Study in Princeton in 1935.

For Einstein, the concepts of physics are inventions, motivated by the scientist's belief in the inner harmony of the world and the confidence that it is possible to grasp reality through theoretical and mathematical constructions. The apparent failure of causality was a failure of human invention, not a failure in the natural world.

Particles, Energy, and Nuclear Science in the 1930s

By the 1930s, after a decade in which the quantum mechanics of electrons had dominated their attention, many physicists, like Fermi in Rome, were convinced that a new frontier lay beyond atomic physics, in the physics of the nucleus. "The problem of equipping the Institute for nuclear work is certainly becoming ever more urgent if we do not want to fall into a state of intellectual slumber," Fermi wrote in September 1932 to Emilio Segrè (1905–1989), who was spending his Rockefeller fellowship year in Germany and Holland.[26]

Eddington and others called 1932 an *annus mirabilis* (miraculous year) in experimental physics. It was the year that Abraham Flexner, founder of the Institute for Advanced Study at Princeton, proposed that Albert Einstein leave Europe and become a permanent member of the institute. In Cambridge, James Chadwick (1891–1974), following up on a report by Irène Curie (1897–1956) and Frédéric Joliot, identified the neutral nuclear particle that Ernest Rutherford had been predicting since 1920. And at Columbia University, Harold Urey (1893–1981) isolated the heavy isotope of hydrogen that became known as deuterium.

On the other side of the United States, in Pasadena, California, Carl Anderson (1905–1991) saw the track in a cloud chamber of a particle that he realized had to be a positively charged electron. P. M. S Blackett (1897–1974) and G. P. S. Occhialini (b. 1907), working together at the Cavendish Laboratory, immediately obtained even better photographs than Anderson had been able to. They correctly interpreted the positively charged particle as an empirical confirmation of Dirac's little-understood theoretical hypothesis of negative energy states or "anti-electrons."

Also working at the Cavendish Laboratory, John D. Cockroft (1897–1967) and Ernest Walton (b. 1903) became the first scientists to split the atom by using artificially accelerated particles, bombarding lithium with protons and disintegrating the lithium nucleus into two helium nuclei. Six thousand miles away, M. Stanley Livingston (1905–1986) and Ernest O. Lawrence (1901–1958) chalked up a symbolic milestone when they achieved an energy of 1 million electron volts with their eleven-inch circular particle accelerator (the cyclotron) at Berkeley.

Why a circular accelerator? The effort to build a particle accelerator followed from frustration with the natural limit on the energy of alpha particles emitted by samples of radium or polonium. A large amount of radium—a gram, for instance—emits some 37 billion alpha particles per second with average energies of 1 to 3 million electron volts. A particularly strong polonium sample may average 5 million electron volts. But in order for charged particles to penetrate the electric field of a target nucleus of a heavy atom, much larger amounts of energy are necessary.

Energy calculations for particle capture by a nucleus were crucial in the laboratory work. Indeed, the key to Chadwick's discovery in Cambridge of the long-sought neutron was his skepticism about the Joliot-Curie report that naturally occurring alpha particles from polonium, when targeted on the light element beryllium, produced gamma rays that could eject protons from paraffin. It turned out that Chadwick was correct in thinking that the alleged gamma rays, while neutral in charge, had to be particles with mass sufficiently high to knock protons out of a paraffin sheet.

Cockroft and Walton at the Cavendish Laboratory, Robert Van de Graaf at MIT, and Merle Tuve (1901–1982) at the Carnegie Institution of Washington were among the physicists and engineers working to build high-voltage, high-tension accelerators in the 1920s and early 1930s. Ernest Lawrence, who, along with his friend Tuve, had studied physics at the University of Minnesota before completing a Ph.D. at Yale, was interested in this problem when he joined the physics department of the University of California at Berkeley in 1928.

In 1929 Lawrence noticed a description by Rolf Wideröe, a Norwegian engineer working in Germany, of a linear device that repeatedly used low and alternating voltages to accelerate charged particles. This was the kind of arrangement used by Cockroft and Walton to accelerate protons. It occurred to Lawrence to use a circular track rather than a linear one, so that particles could be indefinitely accelerated, moving faster and in ever widening circles after each voltage kick. He was confident of success, enthusiastically telling friends, "I'm going to bombard and break up atoms!" "I'm going to be famous!"[27]

While Lawrence was right, it was in fact Cockroft and Walton were the ones who first announced, a few weeks after Chadwick's discovery of the neutron, the disintegration of a light lithium target with a proton beam from their low-voltage accelerator. Nonetheless, the cyclotron design gradually made its way during the late 1930s into laboratories throughout the United States and Great Britain, as well as into the institutes of Bohr, Joliot, and of Karl Manne Siegbahn (1886–1978) in Stockholm. Japanese physicists were among those who came to Berkeley and went home with every detail firmly in mind.

While Lawrence concentrated on improving his machines, diversifying their uses for chemical and medical applications and striving for higher and higher energies at ever-greater expense, the use of accelerating particles from naturally radioactive sources could still produce astonishing results. By 1934, the husband-and-wife team of Frédéric Joliot and Irène Curie, who had begun signing some of their papers jointly as Joliot-Curie in 1928, triumphantly reported the capture of alpha particles by aluminum, followed by emission of neutrons and creation of a new radioactive isotope of phosphorus.

This work, accomplished at the Radium Institute, was done by irradiating an aluminum foil atop a very strong polonium preparation. They found that the foil remained radioactive when the polonium was removed, and they proposed that the aluminum nuclei were absorbing alpha particles, thus becoming heavier and emitting neutrons. Despite the initial denial by Lise Meitner in Berlin that neutrons were emitted, the Joliot-Curie team calculated that the nuclear transformation must be creating an artificial isotope of phosphorus, and they found a way to separate the phosphorus from the aluminum in the astonishingly short time of three minutes.

Showing the test tube of phosphorus that they had isolated to Marie Curie, her daughter and son-in-law relished the older scientist's pleasure as she held the tube in front of a geiger counter. "I will never forget the expression of intense joy that seized her," Joliot said later. Marie Curie wrote in a letter, "We have returned now to the glorious days of the old laboratory."[28] A few months later she died of leukemia.

In the 1930s, laboratory work for the Joliot-Curies and many other scientists was not only punctuated by professional conferences, like the Solvay Physics Congress in Brussels in 1933, where discussions centered on the atomic nucleus, but was also interrupted by an increasing number of political meetings and rallies. In France the instability of national governments in combination with right-wing admiration for Italian and German fascism led to antiparliamentary riots in 1934. Langevin, Perrin, and Joliot, all of whom had long been involved in human- rights and pacifist organizations, now led the organization of an antifascist committee of scholars and intellectuals.

Following the signing of the Franco-Soviet Pact in 1935, a unified front was forged among French socialists, communists, and the centrist Radicals, resulting in an election victory for the Popular Front under Léon Blum in the spring of 1936. Irène Joliot-Curie and then Perrin served in Blum's short-lived government in a cabinet post for scientific research (this in a country where women did not gain suffrage until 1946). But the events of the civil war in Spain, which had begun in 1936, and the German army's occupation of the Rhineland in 1936 soon hopelessly divided the French left, well before the Munich agreement in September of 1938 and the Soviet-Nazi Pact of August 1939.

As the Joliot-Curies found themselves increasingly unable to ignore politics in Paris, the same was becoming true in a rival laboratory in Rome. In the early 1930s, Fermi was well established at the University of Rome as the director of an important laboratory whose staff included Franco Rasetti (who had recently returned from a Rockefeller fellowship year at Caltech and Berkeley), Ettore Majorana, Edoardo Amaldi, Bruno Pontecorvo, and Segrè.

Following his education in Pisa, Fermi had spent time in Göttingen, where he became acquainted with Born, Heisenberg, and Pauli, and in Leyden, where he had worked with Ehrenfest. By the late 1920s, Fermi enjoyed an international reputation for his development of a statistical method (the "Fermi statistics") for predicting the properties of electrons behaving according to Pauli's exclusion principle. In the early 1930s he successfully explained apparent anomalies in radioactive beta-decay with a theory of the emission of electrons paired with massless neutral particles smaller than Chadwick's "neutron": the *neutrino* (Italian for "little neutron").

Physicists and chemists at the Cavendish Laboratory, the Radium Institute, Fermi's institute, and elsewhere realized that the neutron particle, with a mass comparable to the proton but with no charge, could be an extraordinarily useful projectile for exploring and transforming the atomic nucleus. Neutrons were a product of the artificial radioactivity first demonstrated by the Joliot-Curies. They were used by Fermi, working with a radon-beryllium source, to irradiate a series of elements of increasing atomic number. His laboratory produced approximately forty new radioactive substances in a few years' time.

Some pressure was exerted on Fermi from the Italian Academy and the fascist government to name any new elements in honor of the new political order, for example "Littorio" for the lictors, the ancient Roman officers who bore the fasces—the symbol of authority appropriated by Mussolini's fascist government. Mario Corbino, head of the physics institute at the University of Rome and a former government minister, quipped that since the new elements had such very short lives, they might not be appropriate for memorializing fascism.[29]

In 1934 members of Fermi's group made two significant discoveries. They found that a complicated radioactivity was associated with the irradiation of the heaviest known element, uranium (atomic weight 238 and atomic number 92). They concluded that neutron capture by the uranium nucleus might be occurring, followed by beta decay and the creation of a transuranic element of atomic weight 239 and atomic number 93.

They also discovered, "both by chance and by good observation" as Segrè put it,[30] that neutrons filtered through paraffin were considerably more effective in producing nuclear reactions than were neutrons emerging directly from the original source. To test the hypothesis that hydrogen nuclei were what were slowing down the neutrons, the physicists transported their apparatus to the pond in Corbino's garden and found that the water of the goldfish pond worked just as well as paraffin.

Their personal contact with other groups, as well as their publications, led to repetition of the Rome team's experiments and criticism of its results. Irène Curie irradiated thorium and studied the results, using lanthanum as a chemical carrier in solution. She found that the lanthanum precipitate was strongly radioactive. She speculated that a new isotope chemically similar to lanthanum, perhaps an isotope of actinium (atomic number 89), had been created.

At the University of Freiburg the chemists Ida Tacke Noddack (1896–1978) and Walter Noddack (1893–1960) were considerably more skeptical about Fermi's conclusions. They were experts on the rare earths and were interested in the possible discovery of natural transuranic elements. Ida Noddack published an article in 1934 arguing that the Italians should carry out systematic chemical tests to eliminate the possibility that rare-earth elements of lower atomic weight than uranium were present:

> It would be equally possible to assume that when a nucleus is demolished in this novel way by neutrons nuclear reactions occur which may differ considerably from those hitherto observed in the effects produced on atomic nuclei by proton and alpha rays. It would be conceivable than when heavy nuclei are bombarded with neutrons *the nuclei in question might break into a number of larger pieces* which would no doubt be isotopes of known elements but not neighbors of the elements subjected to radiation.[31]

Fermi found it impossible to imagine that atomic numbers were changing by more than one or two units, and Otto Hahn (1879–1968) and Lise Meitner at the Kaiser Wilhelm Institute in Berlin agreed with him.

Within a couple of years the Rome group began disbanding. Rasetti returned to the United States and Pontecorvo joined Joliot in Paris. The formation of the Rome-Berlin Axis in 1936 and the institution of anti-Semitic laws in Italy posed a real threat to Fermi, whose wife, Laura Capon, was the daughter of a Jewish officer in the Italian navy. As mentioned earlier, the couple left Italy in the winter of 1938 when he was given permission to travel to Stockholm to accept the Nobel Prize. By this time the physicist Meitner, who had been collaborating with the chemist Hahn in studies of radioactivity and radiochemistry, had left Germany for Sweden. Her nephew Otto Frisch (1904–1979) had also departed for a safer post with Bohr in Copenhagen.

Hahn's new collaborator, Fritz Strassmann (1902–1980), was considerably less suspicious of the work going on in the Radium Institute than Meitner had been. He paid close attention to a 1938 paper by Irène Curie and her Yugoslav collaborator Pavle Savitch in which they claimed that a radioactive isotope had been produced that chemically followed through in precipitation with lanthanum. Strassman persuaded Hahn that they might have missed something earlier. And so they had. In an article sent off to the journal *Naturwissenschaften* in December 1938 they reported,

> As a consequence of these investigations we must change the names of the substances mentioned in our previous dis-

> integration schemes, and call what we previously called radium, actinium, and thorium, by the names barium, lanthanum, and cerium. As nuclear chemists who are close to the physicists, we are reluctant to take this step that contradicts all previous experiences of nuclear physics.[32]

It was not something *chemically similar* to lanthanum or barium, but lanthanum and barium themselves that were the precipitates. The uranium nucleus had been split into smaller atoms. The paper, published 6 January 1939, arrived in Joliot's office by mail about ten days later. Almost at the same time, the Princeton University physics department's journal club heard the news from the Belgian physicist Léon Rosenfeld (1904–1974), who had just accompanied Bohr across the Atlantic.

As he had been embarking for the United States in January, Bohr learned of Hahn and Strassman's work from Frisch, who was with his aunt Lise Meitner when she received Hahn's letter about the discovery. "Very gradually," Frisch wrote later,

> we realized that the break-up of a uranium nucleus into two almost equal parts . . . had to be pictured [as] . . . the gradual deformation of the original uranium nucleus, its elongation, formation of a waist and finally separation of the two halves. . . . The most striking feature of this novel form of nuclear reaction was the large energy liberated. . . . But the really important question was whether neutrons were liberated in the process, and that was a point which I, for one, completely missed.[33]

Frisch gave Bohr part of a draft of his and Meitner's preliminary theory of what they soon called atomic fission, and during their trans-Atlantic crossing Bohr and Rosenfeld discussed how to interpret this discovery.

The news spread quickly among theoretical physicists in the Princeton and New York area. Eugene Wigner, a Hungarian refugee teaching at Princeton, told Leo Szilard, another Hungarian refugee, who had come down to Princeton from New York for a visit. I. I. Rabi (1898–1988), an emigré physicist teaching at Columbia University, was in Princeton that week and returned to Manhattan to tell Fermi. Szilard was also eager to talk with Fermi: "I thought that if neutrons are in fact emitted in fission, this fact should be kept secret from the Germans."[34]

In the meantime, Frisch had carried out experiments in Copenhagen that detected highly charged and energetic fragments emitted after neutron irradiation of uranium. He found that the effects doubled when he wrapped his neutron sources in paraffin. Frisch asked William Arnold, a California microbiologist visiting Bohr's institute on a Rockefeller fellowship, what biologists called the division of a bacterium into two parts. "Fission" was the answer, and so the word entered the language of nuclear physics in two papers Frisch sent off for publication on 17 January 1939 to the widely read English journal *Nature*.

Some fifty participants in a late-January conference on theoretical physics in Washington, D.C., heard Bohr and Fermi discuss the splitting of the atom. Within a few days newspaper reports reached the west coast. Luis Alvarez (1911–1988) later recalled Robert Oppenheimer's reaction:

> I remember telling Robert Oppenheimer that we were going to look for [ionization pulses from fission] and he said, "That's impossible" and gave a lot of theoretical reasons why fission couldn't really happen. When I invited him over to look at the oscilloscope later, when we saw the big pulses, I would say that in less than fifteen minutes Robert had decided that this was indeed a real effect and . . . he had decided that some neutrons would probably boil off in the reaction and that you could make bombs and generate power, all inside of a few minutes.[35]

Scientists and the Outbreak of World War II

Joliot's laboratory was one of a dozen or so that began concentrating on the uranium problem in 1939. Once he was appointed to the Collège de France in 1937, Joliot had been able to build a laboratory dedicated to nuclear physics; and funds from the newly established Centre Nationale des Recherches Scientifiques helped support some postdoctoral researchers. One of these was Hans Halban (1908–1964), an Austrian of Jewish descent whose study of nuclear physics included a year in Copenhagen working with Frisch. Another was Lew Kowarski (1907–1979), a Russian of Jewish descent and a chemical engineer who finished his doctoral thesis in Perrin's laboratory for physical chemistry. Following Hahn and Strassman's reports on barium and lanthanum, Joliot and his new team immediately set out to detect signs of neutron production.

The idea of a chain reaction was not novel in 1939. Joliot, in his 1935 Nobel Prize address, had suggested that energy from the atomic nucleus might one day be liberated in a chain reaction of atomic transformations taking place one after another. Indeed in early 1934 Szilard, who had just emigrated to England, applied in London for a patent for a process of a nuclear chain reaction.

An avid reader of H. G. Wells's science fiction, Szilard had been deeply impressed by Wells's 1913 novel *The World Set Free,* a tale of devastating war in which "atomic bombs" destroy hundreds of cities. After this disaster, atomic energy is used by survivors in desert and arctic wastelands to turn what is left of the world into lush gardens and prosperous cities.

Szilard became fascinated with the idea of atomic energy. Following the announcement of Chadwick's discovery of the neutron, Szilard predicted that an atom might be made artificially radioactive, releasing huge stores of energy if it could be bombarded with Chadwick's neutral particles. Szilard tried to interest British scientists and the British admiralty in his ideas, advising them, too, that this kind of knowledge was sufficiently dangerous that it should be kept secret. They paid him little attention.

By the end of February 1939, Kowarski, Halban, and Joliot were ready to publish a joint paper confirming that the fission of uranium produces neutrons. Their conclusion was based on investigations with a low-energy radium-beryllium source sunk in a tank of water in which a uranium compound was dissolved. They decided to publish this paper in *Nature* so that it would get the widest possible audience among scientists. As Halban later said, "We were thoroughly convinced that the conditions for establishing a divergent chain reaction with neutrons could be realized."[36]

As the French team was carrying out these investigations during February 1939, a letter arrived in Joliot's office from Szilard suggesting that if Joliot, like Fermi at Columbia, were looking for neutron emission, it might be best to avoid publication of the results. Szilard simultaneously approached other scientists about setting up a code of secrecy. He enlisted Victor Weisskopf, then at Princeton, to write Blackett, and Szilard asked Wigner to write P. A. M. Dirac about the matter. Szilard also tried to persuade Columbia University physicists not to send their results to *Physical Review.* His effort failed when it became clear that Joliot's team had published in *Nature* and intended to keep on publishing.

Germany had by now invaded the rest of Czechoslovakia and closed the uranium mines of Joachimsthal (Jáchymov) to foreign exports. In July 1939, one of Hahn's Berlin colleagues, Siegfried Flügge, wrote an account of the uranium chain reaction for *Naturwissenschaften* and discussed it in an interview with a journalist from the *Deutsche Allgemeine Zeitung*.

Together, Szilard and Wigner urged Einstein to call on his friendship with Queen Elizabeth of Belgium to guarantee that export to Germany of uranium ores in what was then the Belgian Congo (now Zaire) would be prohibited. Einstein also agreed to write what became a famous letter to President Franklin D. Roosevelt, a letter delivered to Roosevelt three months later by the banker and scholar Alexander Sachs. The outcome was the formation of the Advisory Committee on Uranium, although it accomplished little in the course of the next year.

That same summer of 1939 Heisenberg made an extended trip to the United States, including a visit to Ann Arbor, Michigan, where he stayed with Goudsmit for the annual Ann Arbor summer school of physics. Heisenberg had just turned down the offer of a faculty position at Columbia University after receiving personal assurance from the head of the German police forces, Heinrich Himmler, that verbal attacks on Heisenberg orchestrated by the SS would cease. Enrico and Laura Fermi, along with Goudsmit, encouraged Heisenberg to reconsider his decision to stay in Germany, but he replied that the excesses of the Hitler regime were bound to stop and that the physics community in Germany needed him.

Heisenberg later recalled that he and Fermi talked in Ann Arbor about Fermi's expectation that after an outbreak of war, scientists in all nations would "be expected by their respective governments to devote all their energies to building the new weapons." But Heisenberg recalled that he replied to Fermi that "for the present I believe that the war will be over long before the first atom bomb is built."[37] There was no talk of keeping atomic fission secret among the physicists.

Just over a month later, on 3 September 1939, two days after the German army invaded Poland, England and France declared war on Germany. On 30 October 1939 Joliot and his colleagues deposited a sealed note at the Paris academy recording theoretical calculations for a nuclear chain reactor. There the note remained until 1948, and the Joliot laboratory published nothing further on nuclear research until the war was over.

By the end of 1939, as estimated by the historian Spencer Weart, some one-hundred papers on fission had been published in western Europe and the United States, 40 percent of them in the United States, 25 percent in France, 15 percent in Germany, and 10 percent in Great Britain. Nuclear research in the Soviet Union included work in the laboratory of physicist Igor V. Kurchatov (1903–1960), whose associates reported to the *Physical Review* in June 1940 that they had observed rare spontaneous fissioning in uranium. There was no American response to the paper.

In Japan, the army research engineer Takeo Yasuda noted published reports on fission and assigned the problem to Yoshio Nishina (1890–1951) who had studied with Bohr and was building his second cyclotron at his Tokyo laboratory. The Imperial Army Air Force, confident of uranium supplies in Korea and Burma, authorized atomic bomb development in early April 1941. The elderly Hantaro Nagaoka served on a scientific committee for the bomb along with Nishina.

In April 1940, German troops occupied Denmark and Norwegian ports, in May Germany invaded the Netherlands and Belgium; and in June Germany defeated France. Joliot's stocks of heavy water were smuggled to England by Halban and Kowarski, where by July 1940 considerable progress had been made on studies of the nuclear chain reaction.

This work was headed by George P. Thomson who gave his top secret committee of British scientists the meaningless name MAUD. A technical subcommittee included the foreign emigrés—Halban, Kowarski, Frisch, and Rudolf Peierls (b. 1907). For the most part they worked in Cambridge.

While Bohr was in Princeton during 1939, he figured out in collaboration with John Wheeler (b. 1911) that it is uranium 235–a rare isotope—and not the common uranium 238 that is fissionable by slow neutrons. The MAUD group estimated that the minimum amount of the fissionable isotope of uranium 235 required for a self-sustaining chain reaction was only ten kilograms, astoundingly far below the figure of thirteen tons that had recently been calculated by Francis Perrin (1900–1992), and the figure of several tons first calculated by Peierls.

The British group was confident that gaseous diffusion could be used to separate U-235 from U-238 in natural uranium compounds. They also noted the significance of the report by Berkeley cyclotroneers Edwin McMillan and Philip Abelson that accelerated neutrons could transform U-238 into fissionable plutonium.

By the summer of 1940 many British scientists, including Blackett, became involved with the war effort for improvements in radar and the analysis of military equipment, weapons, and tactical operations. This was a cooperative effort between the Royal Air Force and the Air Ministry, which appointed scientists to study the defense of Great Britain against German air attacks. Operations research was so successful that it was extended from air to naval operations in response to attacks by German U-boats on trans-Atlantic traffic between Great Britain and the United States.

During the summer of 1940 the National Defense Research Committee (NDRC) was established in the United States under the leadership of Vannevar Bush (1890–1974), previously a professor of power transmission at MIT and then president of the Carnegie Institution in Washington, D.C. Bush argued that scientists must be mobilized directly under a federal agency with federal funding. Working through President Roosevelt's uncle Frederick Delano, who was a trustee of the Carnegie Institution, as well as presidential adviser Harry Hopkins, Bush succeeded in persuading Roosevelt to establish an agency by executive order.

The NDRC included James B. Conant (1893–1978), a chemist and the president of Harvard University; Karl T. Compton (1887–1954), a physicist and the president of MIT; Frank B. Jewett (1879–1949), the director of Bell Labs and president of the National Academy of Sciences; and Richard C. Tolman (1881–1948), a physical chemist from Caltech. The committee's mandate was to "correlate and support scientific research on the mechanisms and devices of warfare, except those relating to problems of flight," by entering into contracts with individuals and institutions both private and public.[38]

After a year, a new agency incorporated the NDRC and a medical division under the title Office of Scientific Research and Development (OSRD). By the end of the 1945–1946 fiscal year, this agency was to expend more than $453 million, with enormous amounts of contract money going to MIT, Caltech, Harvard, Columbia, and the University of California as well as to Western Electric, Du Pont, RCA, Eastman Kodak, and General Electric. The "Rad Lab" at MIT for microwave radar research and development, a Caltech project for solid fuel rockets, and a Johns Hopkins University facility for development of the proximity fuse (for exploding artillery shells) were the largest projects partly funded by OSRD. Indeed, as the historian Daniel J. Kevles has noted, the combined total costs for these three projects far exceeded the $2 billion spent on the more famous Manhattan Project.

On Saturday, 6 December 1941, Bush and Conant called together members of the Uranium Committee for a weekend meeting in Washington, D.C., to reorganize and revitalize their work. Harold Urey was to develop gaseous diffusion at Columbia University, Ernest Lawrence would undertake electromagnetic separation of uranium isotopes at Berkeley; and Eger V. Murphree, the director of research for Standard Oil of New Jersey, would supervise centrifuge development. Arthur Holly Compton (1892–1962) would be responsible for theoretical studies and design of the bomb at the University of Chicago, a project that would focus on both uranium and plutonium. On Sunday, 7 December, the Japanese attacked the U.S. naval installation at Pearl Harbor, Hawaii, and the United States was suddenly at war with Japan, Germany, and Italy.

Reaction to the declaration of war among American chemists and physicists, as in the larger scientific and engineering community, was largely one of support. As recently as the winter of 1938, some thirteen hundred American scholars and scientists had signed their names to a manifesto that condemned Nazi racial theories, asserted the legitimacy of theoretical physics in Germany, and defended freedom of thought in all spheres.

Some of these scientists were members of a delegation from the American Association for the Advancement of Science to the fall 1938 meeting of the British Association for the Advancement of Science in Cambridge. The questions that these scientists discussed at the meeting may sound abstract, but they carried immediate practical and moral meaning in 1938: Were contemporary social upheavals due to the advance of science and technology? Is freedom of thought necessary for the progress of science? How can scientists help to preserve freedom and protect science and themselves?[39]

Many American scientists, like their fellow American citizens across all occupations and professions, had been isolationists in the early 1930s. As late as the winter of 1939–1940, after the Germans had marched into Poland, a pacifist petition arguing for U.S. neutrality and denouncing the military use of science was circulated by leftist members of the Boston-area American Association of Scientific Workers.

American scientists, by the nature of their international contacts and friendships, were often personally aware of the limitations on freedom and the militarism that were coming to dominate Central Europe during the decades of the 1920s and 1930s. Thousands of Jewish and leftist men and women, many of them highly trained in professional fields, had been forced to leave Europe since 1933 or had chosen to do so earlier. Many of those who had stayed were in peril of their lives, and for those who were not, the cultural and professional values they shared with foreign colleagues were in danger of elimination in the new National Socialist order of the Third Reich.

By the autumn of 1940, following the defeat of France, thousands of American scientists and engineers responded to a presidentially mandated questionnaire from the National Resources Planning Board to establish a national roster of technical personnel for projects critical in time of war. One of those who responded was Albert Einstein. He wrote, "I was always interested in practical technical problems."[40]

Whether they preferred the university lecture hall, the library, the laboratory, the governmental agency, or the administrative bureaucracy, chemists and physicists of Einstein's generation had found themselves pursuing an internationalist program of knowledge during decades of intense civil unrest, national rivalry, and brutal killing and war. For many of these scientists, as for many of their predecessors in earlier centuries, their search for scientific understanding was an everyday matter of work, career, and useful results, but it also was a spiritual quest that from time to time lifted them above the battle. By 1940 it was clear that there was no escape from the battle. It was in this milieu that nuclear science had been created.

Conclusion: The Claims of Science

After 1940 science came to be more closely dependent on industry and government, particularly in military-related research, than at any time in its past. The Soviet Union's launching of *Sputnik* in 1957 was a stunning event that immediately led to increased emphasis in the United States and elsewhere on scientific education in the physical sciences and engineering, as well as on research that could lead to new generations of accelerators, missiles, and space satellites. The politics of the Cold War created "big science" projects that in the United States required Congressional approval and popular support for enormous expenditures of money not only in scientific areas related directly to defense, but also health, energy, and agriculture.

The science of the Cold War also demanded many scientists' acquiescence to a patriotism of military and industrial secrecy that ran counter to the values of open cooperation and competition that scientists traditionally believed should characterize their work.[1] Some scientists' attitudes on nuclear weapons research and disarmament brought them into conflict with their governments in the early years of the Cold War, most spectacularly in the case of Robert Oppenheimer. Many scientists feared that they could not sustain their independence in the techno-scientific structure of postwar research.

Before 1940 the growth and proliferation of scientific disciplines and subdisciplines had been based largely in the scientific activity of western Europe. After World War II, the development of scientific and technological (techno-scientific) infrastructures took place in countries throughout the world, many of them newly independent from the administrative and educational control of the western European and North American powers. The numbers of women scientists and engineers gradually increased. Thus, the structure of the scientific enterprise became even more complex and diverse after 1940.

For all this seemingly radical change, much remained continuous with the past. Most of the basic presuppositions and fundamental explanatory systems in chemistry and physics changed very little from those of the prewar period, even while postwar instrumentation, much of it computer-driven, became bigger, faster, and more complex.

Some of the greatest achievements of nineteenth and early twentieth-century chemistry and physics—the ideas informing the framework for Mendeleyev's periodic table of the elements, the structural theory of organic chemistry, the laws of thermodynamics, the electron theory of chemical bonding and reaction, the special and general theories of relativity, the basic assumptions of quantum mechanics and nuclear science—remained guideposts for late twentieth-century physical science.

Chemistry and physics continued to be defined and taught as different disciplines, although the problems and practices of chemistry and physics overlapped more than ever before. A woman's or man's identity as a chemist or physicist had to do with education, apprenticeship, and disciplinary networks, as well as with the choice of problems, the means of investigation, and the language and symbols of explanation.

Something that did seem to change in the second half of the twentieth century was the conceptualization among many academic physical scientists of their public role. Until the 1930s many chemists and physicists (or chemical philosophers and natural philosophers) thought of themselves first of all as *intellectuals* and only secondarily as *scientists*. For the most part, their education and interests were broadly gauged and wide-ranging, often encompassing geometry and algebraic mathematics, rhetoric and philosophy, the classics of ancient Greece and Rome, religious thought, and the secular literature of the seventeenth and eighteenth centuries. Many men and women who like Michael Faraday were educated outside the universities admired in science precisely the virtues of a moral and an intellectual vocation. These men and women were not simply scientific "experts" but rather members of a philosophical culture.

The historian of physics Paul Forman is one of those who has argued that there is a difference in the self-identity of physicists before and after World War II. The physics community, he suggests, traded a prewar professional ethos based in the ideal of the physicist as the upholder of moral values for a standard that lionized the physicist as a genius and a "playboy" who was just having "fun."[2] Forman's argument is too general to be entirely true, although it does reveal the prevailing male clubbiness of the physical sciences after the war.

The physicist Richard Feynman (1918–1988) epitomized the playboy image Forman has in mind. Feynman was fourteen years younger than Oppenheimer. They were scientists of two different generations. It was the 1940s and the work of the Manhattan Project that formulated Feynman's worldview, not, as with Oppenheimer, the 1920s and the gentleman's education at Harvard, Cambridge, and Göttingen.

Feynman, a graduate of MIT, was already a staff member of the atomic-bomb project while studying for his doctorate at Princeton University. He later wrote of his education,

> "I didn't have time, and I didn't have much patience, to learn what's called the humanities. Even though there were humanities courses in the university that you had to take in order to graduate, I tried my best to avoid them. It's only afterwards, when I've gotten older and more relaxed, that I've spread out a little bit. I've learned to draw and I read a little bit, but I'm really still a very one-sided person. . . .[3]

Feynman's self-ruminations, like other aspects of the self-portrait he sketched in books such as *"Surely You're Joking, Mr. Feynman"* (1985) and *"What Do **You** Care What Other People Think?"* (1988), create an image that is very different from that of the eighteenth, nineteenth, or early twentieth-century *savant*. This is not to say that chemists and physicists in earlier centuries were all morally virtuous and intellectually cultivated, but for the most part they respected and valued the idea of science as a higher calling to knowledge and truth.

In contrast, Feynman's books exemplify a postwar, postmodern perception that science and scientists have no special calling, no higher responsibility, and no privileged pathway to understanding. The link between science and transcendance has been broken.

Yet for Feynman, as for his scientific predecessors and the scientific public, the sciences "enable us to do all kinds of things and to make all kinds of things." This is part of the practical value of chemistry and physics and other sciences. This theme is one that strikes a chord with traditional statements about the value of science. So does a second theme resonate with earlier generations' presumptions about the value of science. Science, Feynman argued, teaches us about freedom and responsibility. And this is not just the "playboy's freedom":

> Scientists are [aware]. . . . that it is possible to live and *not* to know. But I don't know whether everyone believes this is true. Our freedom to doubt was born out of a struggle against authority in the early days of science. It was a very deep and strong struggle. . . . It is our responsibility as scientists. . . . to proclaim the value of this freedom; to teach how doubt is not to be feared but welcomed and discussed; and to demand this freedom as our duty to all coming generations.[4]

Feynman here highlights one of the paradoxes of the pursuit of science: its quest for "certainty" and its method of "doubt." Descartes knew this, of course, as did Galileo.

Perhaps no one has made a stronger case for rationality and skepticism than the philosopher Karl Popper (1902–1994), who stressed as a criterion for any scientific theory its potential "falsifiability." Nothing is true if it cannot conceivably be proved false. Popper, like many scientists past and present, linked scientific thinking with *clear* and with *rational* thinking. Many late twentieth-century scientists have agreed with Popper on all these scores.

In 1964 Americans ranked nuclear physicists third in occupational prestige after Supreme Court justices and physicians.[5] It is not surprising that, following World War II, nuclear physicists were thought to play as powerful a role in matters of life and death affecting ordinary people as do the High Court's judges and healers of the human body.

The attribution of power was not, however, the only reason that representatives of law, medicine, and science were held in equally great awe. All these professions were traditionally identified with honorable vocations, not avaricious occupations. All carried the expectation of knowledgeable expertise and clear judgment on problems demanding a high degree of certainty for their solution. All were professions that emphasized the roles of observation, reason, investigation, and doubt in ascertaining reliable knowledge.

Toward the close of the twentieth century, after the Vietnam War and the end of the Cold War, the American public appeared disenchanted with law, medicine, and science, in part because of their practitioners' increasingly common identification with careerism and money-seeking and in part because of their inevitable failure to meet the high levels of expectation placed in them.

It has been argued by some historians and sociologists of science in the last decade or two that scientists have no special warrant on the truth and that they behave like ordinary people with no privileged insights into problem-solving. Scientists themselves, Feynman among them, have contributed to this substantial demystification of the sciences.

Yet, if many late twentieth-century scientists no longer adhere to the ethos of transcendance in the tradition of their predecessors, most, like Feynman, remain defenders of the method of rationality and skepticism grounded in evidence. Most chemists, physicists, and other scientists also persist in the notion that progress, if not perfection, has been among the achievements of the sciences, resulting in knowledge that is reliable, useful, and remarkably universal in its intellectual appeal and practical implications. These are modest claims for the sciences of chemistry and physics, as we understand their history during the period from 1800 to 1940.

Chronology

1799 The *coup d'état* of Napoleon Bonaparte

1800 The Royal Institution in London receives a charter from George III

1808 John Dalton publishes the first part of *A New System of Chemical Philosophy*

1811 Amedeo Avogadro proposes that equal volumes of gases contain equal numbers of molecules under similar conditions

1813 Jöns Jakob Berzelius advances a system of alphabetical symbols for the elementary atoms

1814 Pierre-Sadi Laplace publishes the *Philosophical Essay on Probability*. His *Celestial Mechanics* continues to appear in five volumes from 1798 to 1827

1815 The defeat of Napoleon Bonaparte at Waterloo

William Prout proposes that all elementary atoms are made up of hydrogen atoms

1819 Augustin Fresnel receives the Grand Prize of the Paris Academy of Sciences for his mathematical treatise on the wave characteristics of light

1822 Jean-Baptise-Joseph Fourier's *Analytical Theory of Heat*

1824 Sadi Carnot publishes *Reflections on the Motive Power of Fire*

1825 Michael Faraday becomes director of the Royal Institution laboratory

1826 Justus Liebig establishes chemical laboratory at Giessen

1828 University College London founded with admission policy open to all men regardless of religion

1829 Auguste Comte begins giving course of lectures in positive philosophy, publishing his first volume of lectures the following year

1831 The opening of the Liverpool-Manchester railroad

1832 Justus Liebig establishes the *Annalen der Pharmacie*, the predecessor of *Liebigs Annalen der Chemie*

1837 Victoria becomes Queen of Great Britain and Ireland

Auguste Laurent defends his thesis of the hydrocarbon nucleus

1838 Jean-Baptiste Dumas publishes on the chemical type

1845 Michael Faraday coins the term *field* for describing the transmission of electromagnetic effects

1846 William Thomson assumes chair of natural philosophy at Glasgow University

1847 Hermann von Helmholtz publishes "On the Conservation of Force"

James Joule lectures at BAAS on the mechanical equivalent of heat

1848 Revolutions sweep through Europe

1849 Louis Pasteur discovers optical isomers in tartrate crystals

1851 The natural sciences tripos is established at Cambridge University

William Thomson publishes "On the Dynamical Theory of Heat"

1853 Charles Gerhardt begins publishing the four-volume *Treatise on Organic Chemistry*, completed in 1856

1854 William Thomson coins term *thermodynamics*

1856 William Henry Perkin Sr. prepares the first synthetic dye aniline purple or mauve

1857 Rudolf Clausius publishes "On the Kind of Motion We Call Heat"

1857 August Kekulé sketches a theory of the tetravalence of the carbon atom

1858 August Kekulé and Archibald Scott Couper each publish detailed papers proposing the linking of carbon atoms to form a backbone for the hydrocarbon molecule

1860 Stanislao Cannizzaro delivers his paper "On a Course of Chemical Philosophy" at the international chemistry conference in Karlsruhe

James Clerk Maxwell's "Illustrations of the Dynamical Theory of Gases" establishes statistical laws in mechanics

1861 Civil war begins in the United States of America

Emancipation of the serfs in Russia by Czar Alexander II

James Clerk Maxwell publishes the first two parts of his paper "On Physical Lines of Force"

Aleksandr Butlerov proposes the term *chemical structure* for the constitution of the hydrocarbon molecule

1863 President Abraham Lincoln issues the Emancipation Proclamation

1864 Alexander Crum Brown adopts notation for graphical chemical formulas that uses broken lines for valences

1865 Rudolf Clausius coins the word *entropy*

August Kekulé publishes six-carbon bonds for the structure of benzene

1867 Creation of North German Confederation and election of Reichstag members by universal male suffrage

William Thomson (Lord Kelvin) publishes a widely read paper "On Vortex Atoms" and the *Treatise of Natural Philosophy*, co-authored with Peter Guthrie Tait

1869 Dmitry Mendeleyev publishes the first version of his periodic table of the chemical elements

1871 Defeat of France by Prussia and declaration of the German Empire at Versailles

Completion of Italian unification

James Clerk Maxwell becomes the first director of the Cavendish Laboratory at Cambridge

1874 Jacobus van't Hoff and Joseph-Achille Le Bel each propose the tetrahedron model for carbon valence bonds

1875 William Crookes experiments with vacuum discharge tubes and constructs a radiometer

1876 Edward Alexander Bouchet becomes the first black American to receive a doctoral degree from a university in the United States with a Ph.D. in physics at Yale University

1877 Ludwig Boltzmann publishes a probabilistic interpretation of the second law of thermodynamics

1878 University College London opens admissions to women

1881 Albert A. Michelson and Albert Morley begin a series of experiments to detect optical effects due to the ether

1884 Hermann von Helmholtz becomes director of the newly built Physikalisch Technische Reichsanstalt

Jacobus van't Hoff publishes *Studies in Chemical Dynamics*

Emil Fischer begins investigations of the structure of sugars

1886 Henrich Hertz produces electrical waves 0.3 to 10 meters long in the air

1887 Svante Arrhenius, Wilhelm Ostwald, and Jacobus van't Hoff claim to establish a new discipline of physical chemistry

1888 Joseph John Thomson becomes director of the Cavendish Laboratory

1889 George F. Fitzgerald proposes the hypothesis of contraction of a fast-moving electron in the direction of its motion

1894 Joseph Larmor applies the word *electron* to charged nuclei in the ether

1895 Wilhelm Röntgen discovers X rays

1896 Pieter Zeeman detects the broadening of spectral lines in a magnetic field (the "Zeeman effect")

Gustave LeBon announces a new radiation he calls black light

Henri Becquerel discovers natural radiations emanating from uranium salt

1897 Joseph John Thomson publishes experimental evidence for a universal corpuscle of matter, subsequently renamed the electron

1899 War begins between England and Dutch settlers (Boers) in South Africa

1900 Max Planck proposes the quantum of action to explain experimental energy values for black-body radiation

1901 Death of Queen Victoria

The first Nobel Prizes are awarded

1902 Marie Skłodowska Curie defends her doctoral thesis on radioactivity and the isolation of polonium and radium

Walter Kaufmann reports increase in mass of high-speed electrons

1903 Ernest Rutherford and Frederick Soddy's theory of the transmutation of the nucleus in radioactive atoms

René Blondlot announces a radiation that he calls "N"

Henri Becquerel, Marie Curie, and Pierre Curie share the Nobel Prize in Physics

1905 Japan defeats Russia

Albert Einstein publishes papers on molecular-kinetic theory, the quantum hypothesis applied to light, and the relativity principle applied to the electrodynamics of moving bodies

1906 Walther Nernst states the heat theorem for entropy at absolute zero

1907 Emil Fischer synthesizes an eighteen-unit amino-acid polypeptide in his ongoing studies of protein structure

1908 Marie Curie becomes the first woman professor in the history of the University of Paris

1911 Ernest Rutherford publishes a planetary model for the atom

The first Solvay conference in physics in Brussels

1912 Max von Laue's research group demonstrates the diffraction of X rays by the atomic lattice of a crystal

1913 Henry G. J. Moseley defines atomic number as ordering principle for the periodic table of the elements

Niels Bohr applies the quantum hypothesis to the planetary model of the atom

1914 World War I begins

Outbreak of war prevents test of Einstein's 1911 theory of general relativity during a solar eclipse in Russia

"Manifesto of the Ninety-Three German Intellectuals"

1915 Arnold Sommerfeld publishes *Atomic Structure and Spectral Lines* with proposal of two new quantum numbers

1916 Gilbert Newton Lewis advances theory of the electron-pair valence bond

George Ellery Hale forms National Research Council at President Woodrow Wilson's request

Thomas Alva Edison heads Naval Consulting Board

German troops use chlorine as chemical weapon

Einstein completes a revised and more advanced version of the general theory of relativity

1917 Bolshevik Revolution

Germans introduce mustard gas as chemical weapon

1918 Universal suffrage in Great Britain

1919 Versailles Peace Treaty and founding of the League of Nations

Announcement in London of the confirmation during solar eclipse of Einstein's general theory of relativity

1920 Paul Weyland begins public campaign of vilification against Albert Einstein's scientific work and political views

Nineteenth amendment to the U.S. Constitution gives women the right to vote

1922 Hermann Staudinger and Jakob Fritschi coin the term *macromolecule*

1924 The prince Louis de Brogile defends his thesis on the wave properties associated with the orbital electron

1925 Wolfgang Pauli proposes his principle of exclusion, leading to requirement of fourth quantum number

George Uhlenbeck and Samuel Goudsmit advance idea of electron spin

Werner Heisenberg works out a matrix mechanics for electron energies

1925 United States declines to sign Geneva Convention outlawing use of gas warfare

1926 Erwin Schrödinger reformulates electron wave mechanics

Max Born offers a probabilistic interpretation of Schrödinger's equation

Werner Heisenberg proposes the uncertainty principle

Werner Heisenberg applies mechanical resonance to wave functions of electrons in the helium atom

Edith Hilda Ingold and Christopher K. Ingold suggest there are no true double or single bonds in benzene

1927 Niels Bohr establishes the Copenhagen interpretation of quantum mechanics at conferences at Lake Como and Brussels

Walter Heitler and Fritz London apply quantum wave mechanics to the atomic orbits of electrons in the hydrogen molecule

Friedrich Hund applies quantum wave mechanics to the molecular orbits of electrons in the hydrogen molecule

1929 Stock markets crash and the Great Depression begins

1931 John Slater and Linus Pauling each work out principles of what later is called hybridization of electron energy levels

Japan invades China

1932 James Chadwick identifies the neutron in experiments at the Cavendish laboratory

Carl Anderson and, working independently, P. M. S. Blackett, with G. P. S. Occhialini, each photograph the tracks of the anti-electron, subsequently called the positron

John D. Cockroft and Ernest Walton split an atom of lithium with accelerated protons

Ernest O. Lawrence designs his first cyclotron

1933 Adolf Hitler becomes chancellor of Germany

Civil service laws exclude scientists of Jewish descent from German universities and other civil service positions

Linus Pauling applies resonance hybridization to benzene

1934 Dorothy Crowfoot Hodgkin achieves first x-ray diffraction photograph of a protein (pepsin)

Irène Joliot-Curie and Frédéric Joliot produce radioactivity in naturally non-radioactive phosphorus

Enrico Fermi's group in Rome studies the effects of bombarding uranium with neutrons slowed down in paraffin

1935 Frédéric Joliot's Nobel Prize lecture notes the possibility of atomic energy from a chain reaction

1936 Popular Front achieves political power in France and Irène Joliot-Curie serves briefly as undersecretary of state for scientific research

Civil war begins in Spain

1937 Pact forming Rome-Berlin-Tokyo Axis

1938 Germany invades Austria and the Sudetenland in Czechoslovakia

Munich agreement

Otto Hahn writes Lise Meitner of the splitting of the uranium nucleus by slow neutrons

1939 Germany invades the rest of Czechoslovakia

Frédéric Joliot's group publishes confirmation of neutron emission and a chain reaction associated with uranium fission

Pact between Germany and the Soviet Union

Werner Heisenberg visits the United States and returns to Germany

Germany invades Poland

Great Britain and France declare war on Germany

Linus Pauling publishes *The Nature of the Chemical Bond*

1940 Germany invades Denmark, Norway, the Netherlands, Luxembourg, Belgium, and France

Stocks of heavy water are smuggled from France to England by Hans von Halban and Lew Kowarski

Vannevar Bush becomes head of the new National Defense Research Committee in the United States

Notes and References

Introduction: Modern Science and Big Science

1. See Alvin M. Weinberg, "Impact of Large-Scale Science," *Science*, 134 (1961), 161–64, on 164; quoted in Daniel J. Kevles, *The Physicists: The History of a Scientific Community in Modern America* (Cambridge, Mass.: Harvard University Press, 1987), 418–19.

Chapter 1

1. Letters from Edward Frankland Armstrong to his father, H. E. Armstrong, dated 11 February 1900, and [summer] 1900. The younger Frankland also wrote to his father of increasing anti-Semitism among students in the Berlin laboratories, ascribing it to the view that "nearly all the Berlin chemists are Jews"; letter of 23 December 1900 (Archives of Imperial College, University of London); cited by Mary Jo Nye, *From Chemical Philosophy to Theoretical Chemistry* (1993), 169 and 169 n22.
2. Quoted in Erwin N. Hiebert, "Nernst and Electrochemistry," in *Selected Topics in History of Electrochemistry*, eds. George Dubpernell and J. H. Westbrook(1978), 182.
3. Ludwig Boltzmann, "On the Trip of a German Professor into El Dorado," in *Ludwig Boltzmann*, ed. John Blackmore (1995), 180–81.
4. J. B. Morrell, "The Chemist Breeders," *Ambix* 19 (1972): 1–46.
5. Quoted in Alan J. Rocke, *The Quiet Revolution* (1993), 289.

6. Matthew Arnold, *Schools and Universities on the Continent* (1868), 256.

7. Thomas Sprat, *History of the Royal Society* (1959), 134–38.

8. See G. Norman Burkhardt, "The University of Manchester," *Chemistry and Industry* (1949), 427; quoted in Nye (1993), 164.

9. Alexander Todd, *A Time to Remember* (1983), 44–45.

10. H. E. Armstrong, "Presidential Address," Chemistry Section, *BAAS Reports, Winnipeg (1909)* (1910), 451; quoted in Nye (1993), 167–68.

11. Daniel J. Kevles, *The Physicists* (1995), 211–12.

12. John W. Servos, *Physical Chemistry* (1990), 240 and 245. The Nobel laureates were Harold C. Urey (1933), William F. Giauque (1949), Glenn T. Seaborg (1951), Willard F. Libby (1960), and Melvin Calvin (1961).

13. H. Bence Jones, *The Royal Institution* (1871), 121.

Chapter 2

1. Isaac Newton, *Opticks*, 4th ed. (1730; reprint, 1931), 400.

2. Quoted in Alan J. Rocke, *Chemical Atomism* (1984), 57.

3. From John Dalton, *Memoirs of the Literary and Philosophical Society of Manchester*, [2], 1 (1805), 286; quoted in Aaron J. Ihde, *The Development of Modern Chemistry* (1964), 106.

4. From Jöns Jakob Berzelius, *Essai sur la théorie des proportions chimiques et sur l'influence chimique de l'électricité*, introduced by Colin A. Russell (1819; reprint, New York: Johnson Reprint Corporation, 1972), 23.

5. Quoted in Rocke (1984), 178–79.

6. Quoted in Rocke (1984), 2.

7. Quoted in Christa Jungnickel and Russell McCormmach, *Intellectual Mastery of Nature* (1986), vol. 1, 158.

8. Quoted in Ihde (1964), 229.

9. Auguste Kekulé, "On Some Points of Chemical Philosophy," *Laboratory*, 1 (1867): 303–6, quoted in Ida Freund, *The Study of Chemical Composition* (1968), 624.

Chapter 3

1. William Thomson, "On Vortex Atoms," in *The Question of the Atom*, ed. Mary Jo Nye (1984), 91; see also 92, 94.

2. P. M. S. Blackett, "Memories of Rutherford," in *Rutherford at Manchester*, ed. J. B. Birks (1962), 111–12.

3. Bruce Hunt makes a clear distinction between the approaches of William Thomson and J. J. Thomson, on the one hand, and G. G. Stokes, Oliver Heaviside, and F. G. Fitzgerald, on the other. In contrast, Jed Buchwald and most other historians describe them all as "Maxwellians."

See Bruce J. Hunt, *The Maxwellians* (1991), and Jed Z. Buchwald, *From Maxwell to Microphysics* (1985).

4. In modern terms, 26.8 amps of current running for 1 hour through water releases 1 gram of hydrogen ions and 8 grams of oxygen ions. The total quantity of electricity is 96,400 coulombs, called "1 Faraday" or 1 electrochemical equivalent.

5. Michael Faraday, *Experimental Researches*, vol. 1 (1839), par. 869; quoted in Emilio Segrè, *From Falling Bodies to Radio Waves* (1984), 146.

6. Michael Faraday, *Experimental Researches*, vol. 3 (1855), par. 2146; quoted in Segrè (1984), 149.

7. James Clerk Maxwell, *Elementary Treatise on Electricity* (1881), 52.

8. Crosbie Smith and M. Norton Wise, *Energy and Empire* (1989), 465, quoting from William Thomson's Baltimore lectures; see also 465–67.

9. Pierre Duhem, *The Aim and Structure of Physical Theory* (1954), 70–71.

10. James Clerk Maxwell, "On Faraday's Lines of Force" (1856), in *Scientific Papers of James Clerk Maxwell*, ed. W. D. Niven (1890), vol. 1, 156.

11. Maxwell (1856), in ed. Niven (1890), vol. 1, 208.

12. James Clerk Maxwell, "On Physical Lines of Force," pt. 3 (1862), in ed. Niven (1890), vol. 1, 500.

13. James Clerk Maxwell, *Matter and Motion* (1877), 10; also in *Elementary Treatise* (1881).

14. According to Eugen Goldstein's reminiscences in 1925; quoted in Jungnickel and McCormmach (1986), vol. 2, 89 and 89 n70.

15. Robert S. Cohen, "Hertz's Philosophy of Science: An Introductory Essay" in Heinrich Hertz, *The Principles of Mechanics* (1956), sections 2 and 3.

16. Quoted in William Crookes, "On Radiant Matter," *American Journal of Science* 18 (1879), 242n.

17. Crookes (1879), 261–62.

18. Hermann Helmholtz, "On the Modern Development of Faraday's Conception of Electricity," in *Faraday Lectures*, eds. C. S. Gibson and A. J. Greenaway (1928), 159.

19. J. Norman Lockyer, "Atoms and Molecules," in *Studies in Spectrum Analysis* (1878), 119.

Chapter 4

1. Quoted in Erwin N. Hiebert, "Walther Nernst and the Application of Physics to Chemistry," in *Springs of Scientific Creativity*, ed. Rutherford Aries et al. (1983), 216–17.

2. Quoted from J. G. Crowther, "James Prescott Joule" in Crowther, *Men of Science* (1936), 127–97.

3. Arthur Schuster, *Biographical Fragments* (1932), 201; quoted in Robert Kargon, *Science in Victorian Manchester* (1977), 50.

4. James Prescott Joule, "On Matter, Living Force and Heat" (1847); quoted in A. E. E. McKenzie, *Major Achievements of Science* (1988), 454.

5. William Thomson, *Mathematical and Physical Papers* (1882) vol. 1, 117–18; quoted in Smith and Wise (1989), 315.

6. William Thomson, "On the Mechanical Antecedents of Motion, Heat and Light," *BAAS Report* 24 (1854): 59–63, quoted in P. M. Harman, *Energy, Force, and Matter* (1982), 58.

7. Silvanus P. Thomson, *Life of William Thomson*, vol. 2 (1910), 1143; quoted in Smith and Wise (1989), 108.

8. Robert Clausius, "Le second principe fondamental," *Revue des cours scientifique*, 5 (8 Feb. 1868), 159.

9. Quoted in Rocke (1993), 329

10. Wilhelm Ostwald, "Die Energie und ihre Wandlungen" (1887); reprinted in Ostwald, *Abhandlungen und Vorträge allgemeinen Inhaltes 1887–1903* (1904), 185–206, on 204; quoted in Mi Gyung Kim, "Practice and Representation" (1990), 380.

11. James Clerk Maxwell, "On the Dynamical Evidence," *Journal of the Chemical Society* 28 (1875), 505.

12. James Clerk Maxwell, "A Discourse on Molecules," *Philosophical Magazine* 46 (1873), 466–67.

13. Quoted in Segrè (1984), 242–43.

14. Albert Einstein, *Investigations* (1956), 1–2; this is from his paper "Über die von der molekular-kinetischen Theorie der Wärme Bewegungen von in ruhenden Flüssigkeiten suspendierten Teilchen," in *Annalen der Physik* 17 (1905): 549–60, on 549.

15. Quoted in Mary Jo Nye, *Molecular Reality* (1972), 161–62.

16. Quoted in John L. Heilbron, *The Dilemmas of an Upright Man* (1986), 21.

17. Quoted in Diana Barkan, "The Witches' Sabbath," *Science in Context* 6 (1993), 65.

Chapter 5

1. Caspar Neumann, *The Chemical Works*, abr. ed. (1759), 280, quoted in Frederic L. Holmes, *Eighteenth-Century Chemistry* (1989), 77.

2. Jean-Baptiste-André Dumas, "Mémoire sur la constitution de quelques corps organiques," *Comptes rendus hebdomadaires* (1839): 609–22.

3. Quoted in Ihde (1964), 196.

4. Quoted in Mi Gyung Kim, "The Layers of Chemical Language," *History of Science* 30 (1992), 79; see also David Knight, *The Transcendental Part of Chemistry* (1978).

5. Quoted in O. Theodor Benfey, *From Vital Force to Structural Formula* (1992), 45–46.

6. Quoted in Michael N. Keas, "The Structure and Philosophy of Group Research" (1992), 111.

7. Quoted in Benfey (1992), 8–10.

8. Translated in Benfey (1992), 77.

9. Quoted in Freund (1968), 516–17.

10. Quoted in Freund (1968), e.g., 508, 510.

11. Quoted in Louis F. Fieser and Mary Fieser, *Introduction to Organic Chemistry* (1957), 202.

12. Quoted in Ihde (1964), 326.

13. Rocke (1993), 329.

14. Quoted from Kekulé in Yasu Furukawa, "Hermann Staudinger and the Emergence of the Macromolecular Concept," *Historia Scientiarum* 22 (1982), 3–4; see also Herman F. Mark, "Polymer Chemistry," *Chemical and Engineering News* (6 April 1976): 176–89.

15. Quoted in Judith R. Goodstein, *Millikan's School* (1991), 189.

Chapter 6

1. Quoted in McKenzie (1988), 280.

2. See Robert Reid, *Marie Curie* (1974), 132.

3. [Marie] S. Curie, *Comptes rendus* 126 (1898), 1101 ff.; quoted in Abraham Pais, "Radioactivity's Two Early Puzzles," *Review of Modern Physics* 49 (1977), 930.

4. Marie Curie, in *Revue générale des sciences pures et appliquées* 10 (1899), 41 ff.; quoted in Pais (1977), 930.

5. Quoted in McKenzie (1988), 292.

6. Ernest Rutherford, "Present Problems of Radioactivity," in *Physics for a New Century*, ed. Katherine R. Sopka (1986), 247.

7. Ibid., 248.

8. Alexander Wood, *Cavendish Laboratory* (1946), 23.

9. Quoted in Jun Fudano, "Early X-Ray Research" (1990), 94.

10. J. J. Thomson, "Cathode Rays," *Philosophical Magazine* (1897); reprinted in Nye (1984), 394.

11. Jean Perrin, "Les Hypothèses moléculaires," *Revue scientifique*, 15 (1901): 449–61, on 460; quoted in Mary Jo Nye, *Molecular Reality* (1972), 84.

12. Samuel Glasstone, *Source Book on Atomic Energy*, 2d ed. (New York: Van Nostrand, 1958), 93; quoted in Ihde (1964), 502.

13. Ernest Rutherford, "Scattering of *alpha* and *beta* Particles," *Philosophical Magazine,* series 6, 21 (1911), 624.

14. According to Emilio Segrè, *From X-Rays to Quarks* (1980), 121.

15. See Segrè (1980), 127–28.

16. Quoted in McKenzie (1988), 320.

17. Quoted in Segrè (1980), 140.

18. David C. Cassidy, *Uncertainty* (1992), 197.

19. Quoted in Anatole Abragam, "Louis Victor Pierre Raymond de Broglie," *Biographical Memoirs of Fellows of the Royal Society* 34 (1988), 30.

20. Quoted in Abragam (1988), 29–30.

21. The symbol Δ is sometimes used instead of ∇. ∇^2 is read as "del squared."

22. See Cassidy (1992), 220–21; also Werner Heisenberg, *Encounters with Einstein* (1983).

23. Quoted in Cassidy (1992), 215.

24. Heisenberg (1983), 29.

25. Irving Langmuir, "The Structure of Molecules," *BAAS Reports, Edinburgh (1921)* (1922), 468–69.

26. Robert Mulliken, "Spectroscopy, Molecular Orbitals, and Chemical Bonding," in *Nobel Lectures, Chemistry (1963–1970)* (1972), 137.

27. Linus Pauling, "The Chemical Bond. I." (1931), 1378; quoted in Yuko Abe, "Pauling's Revolutionary Role," *Historia Scientiarum* 20 (1981), 107–24, on 117.

28. Linus Pauling, "The Nature of the Theory of Resonance," in *Perspectives in Organic Chemistry,* ed. Alexander Todd (1962), 7.

29. P. A. M. Dirac, "Quantum Mechanics of Many-Electron Systems," *Proceedings of the Royal Society of London,* series A, 123 (1929), 714.

30. Ralph Fowler, "A Report on Homopolar Valency," in *Chemistry at the Centenary (1931) Meeting of the BAAS* (1932), 226.

31. Linus Pauling Papers, box 242, Popular Scientific Lectures, 1925–1935, quoted in Ana Isabel Simoes, "Converging Trajectories, Diverging Traditions" (1993), 127.

Chapter 7

1. Letters from Edward Frankland Armstrong to his father, H. E. Armstrong, dated 11 February 1900 and [summer] 1900; quoted in Nye (1993), 169.

2. Quoted in Heilbron (1986), 69.

3. Daniel J. Kevles, "Into Hostile Political Camps," *Isis* 62 (1971), 48; for text, see Brigitte Schroeder-Gudehus, *Les scientifiques et la paix* (1978), appendix.

4. Quoted in Thomas P. Hughes, *American Genesis* (1989), 121.

5. Quoted in Hughes (1989), 127.

6. Quoted in Jeffrey Johnson, *The Kaiser's Chemists* (1990), 185–86.

7. R. M. Gattefossé, "L'Industrie chimique à Lyon"; quoted in Mary Jo Nye, *Science in the Provinces* (1986), 189–90.

8. Quoted in Ronald W. Clark, *Einstein* (1972), 280.

9. Quoted in Schroeder-Gudehus (1978), 229.

10. Yasu Furukawa, "Staudinger, Polymers, and Political Struggles," *Chemical Heritage* 20, no. 3 (1993).

11. Quoted in Heilbron (1986), 116.

12. Quoted in Heilbron (1986), 121.

13. Quoted in Heilbron (1986), 155, from an interview with the *New York World Telegram,* in Clark (1972), 458, 462.

14. Robert Jungk, *Brighter than a Thousand Suns* (1958), 37.

15. Quoted in Banesh Hoffmann, *Albert Einstein* (1972), 169–70.

16. "The Fabric of the Universe," *The Times* (London), 7 November 1919, 13; quoted in Clark (1972), 295–97.

17. *The Times* (London), 14 November 1919, 8; quoted in Clark (1972), 295–97.

18. Henri Poincaré, "The Principles of Mathematical Physics," in Sopka, ed. (1986), 281–99.

19. Quoted in Abraham Pais, *Subtle is the Lord* (1982), 148.

20. Albert Einstein and Leopold Infeld, *Evolution of Physics* (1938), 146–47.

21. Quoted in Pais (1982), 178.

22. In Hoffmann (1972), 124–25

23. Arthur S. Eddington, "Sir Frank Dyson, 1868–1939," *Royal Society, Obituary Notices of Fellows* 3 (1940): 159–72, on 167; quoted in John Earman and Clark Glymour, "Relativity and Eclipses" *Historical Studies in the Physical Sciences* 11 (1980), 84.

24. Albert Einstein, "Autobiographical Notes" in *Albert Einstein: Philosopher-Scientist*, ed. Paul Arthur Schlipp, vol. 1 (1959): 2–95, on 81.

25. In Hoffmann (1972), 193.

26. Emilio Segrè, *A Mind Always in Motion* (1993), 86.

27. Quoted in Hughes (1989), 404–5.

28. Quoted in Spencer Weart, *Scientists in Power* (1979), 46, and in Segrè (1984), 199.

29. Segrè (1980), 205.

30. Segrè (1980), 205.

31. Quoted in Jungk (1958), 61.

32. Quoted in Segrè (1980), 207.

33. Jungk (1958), 69–70.

34. Richard Rhodes, *The Making of the Atomic Bomb* (1986), 268.

35. Quoted in Rhodes (1986), 274.

36. Quoted in Weart (1979), 72.

37. From Cassidy (1992), 420, quoting from *Physics and Beyond: Encounters and Conversations* (1971), 170.

38. Quoted in Carroll Pursell, "Science Agencies in World War II," in *The Sciences in the American Context*, ed. Nathan Reingold (1979), 361.

39. In J. G. Crowther, *Fifty Years with Science* (1970), 201.

40. From "Report to the Nation—National Roster of Scientists," NRPB, entry 8, file 106.24; quoted in Kevles (1995), 298; and Pursell (1979), 367–68.

Conclusion

1. Merle A. Tuve, "Is Science Too Big for the Scientist?" *Saturday Review* (6 June 1959), 48–52; summarized in James H. Capshew and Karen A. Rader, "Big Science: Price to the Present," in *Science after '40*, ed. Arnold Thackray, special issue of *Osiris* (7 [1992]) 3–25.

2. Paul Forman, "Social Niche and Self-Image of the American Physicist," in *The Restructuring of Physical Sciences in Europe and the United States, 1945–1960*, eds. Michelangelo de Maria et al. (1989), 96–104; quoted in Capshew and Rader (1992), 12.

3. Richard Feynman,"*What Do **You** Care What Other People Think?*" (1988), 12.

4. Feynman (1988), 240, 245, 248.

5. Kevles (1995), 391.

Bibliographical Essay

Introduction

Among textbooks that are general surveys of the history of science, some include substantial sections on the history of chemistry and physics in the nineteenth and early twentieth centuries. These are Charles C. Gillispie, *The Edge of Objectivity: An Essay in the History of Ideas* (Princeton: Princeton University Press, 1960); A. E. E. McKenzie, *The Major Achievements of Science: The Development of Science from Ancient Times to the Present* (New York: Simon and Schuster, 1960); John Marks, *Science and the Making of the Modern World* (London: Heinemann, 1983); Stephen Mason, *A History of the Sciences: Main Currents of Scientific Thought* (London: Routledge, 1953); and Stephen Toulmin and June Goodfield, *The Architecture of Matter* (New York: Harper and Row, 1962). All have been reprinted since their initial publication.

For excellent introductions and bibliographies to topics in the history of the modern physical sciences, see the splendid book by Stephen G. Brush, *The History of Modern Science: A Guide to the Second Scientific Revolution* (Ames, Iowa: Iowa State University Press, 1988), as well as Pietro Corsi and Paul Weindling, eds., *Information Sources on the History of Science and Medicine* (Boston: Butterworth Scientific, 1983), and R. C. Olby et al., eds., *Companion to the History of Modern Science* (London: Routledge, 1990).

The unexcelled source for biographical studies is C. C. Gillispie, ed., *Dictionary of Scientific Biography*, 16 vols. (New York: Scribner, 1970– 1981), followed by F. L. Holmes, ed., *Dictionary of Scientific Biography,* suppl. 2, 2 vols. (New York: Scribner, 1990). See also Allen G. Debus, ed., *Who's Who in Science: A Biographical Dictionary of Notable Scientists from Antiquity to the Present* (Chicago: Marquis Who's Who, 1968), and *Inventeurs et scientifiques: Dictionnaire de biographies* (Paris: Larousse, 1994). For women scientists, including more extensive biographies for selected women, see Marilyn B. Ogilvie, *Women in Science, Antiquity to the Nineteenth Century: A Biographical Dictionary with Annotated Bibliography* (Cambridge, Mass.: MIT Press, 1986); specifically for women in chemistry and physics, see Louise S. Grinstein, Rose K. Rose, and Miriam H. Rafailovich, eds., *Women in Chemistry and Physics: A Biobibliographic Sourcebook* (Westport, Conn.: Greenwood Press, 1993). For brief biographies of black American scientists, see Vivian Ovelton Sammons, *Blacks in Science and Medicine* (New York: Hemisphere, 1990).

The most inclusive and ongoing bibliography for the history of science is the *Isis Current Bibliography,* published annually as number 5 of the journal *Isis.* Cumulations of this bibliography are Magda Whitrow, ed., *Isis Cumulative Bibliography: A Bibliography of the History of Science Formed from Isis Critical Bibliographies 1–190, 1913–1965,* 6 vols. (London: Mansell, 1971–1984); John Neu, ed., *Isis Cumulative Bibliography 1966–1975,* 2 vols. (London: Mansell, 1980–1983); and John Neu, ed., *Isis Cumulative Bibliography 1976–1985,* 2 vols. (Boston: G. K. Hall, 1989).

For an inventory of archival materials in the history of twentieth-century physics, see J. L. Heilbron and Bruce R. Wheaton, *Literature on the History of Physics in the Twentieth Century* (Berkeley: Office for History of Science and Technology/University of California at Berkeley, 1981), followed by Bruce R. Wheaton, assisted by Robin E. Rider, eds., *Inventory of Sources for History of Twentieth-Century Physics* (Stuttgart: Verlag für Geschichte der Naturwissenschaften und der Technik, 1993).

Selections of original scientific papers in modern chemistry and physics in English translation may be found in O. Theodor Benfey, ed., *Classics in the Theory of Chemical Combination* (1963; rpt., Malabar, Fla.: Robert E. Krieger Publishing, 1981); H. A. Boorse and L. Motz, eds., *The World of the Atom,* 2 vols. (New York: Basic Books, 1966); D. ter Haar, ed., *The Old Quantum Theory: Selected Readings in Physics* (New York: Pergamon, 1967); D. L. Hurd and J. J. Kipling, eds., *The Origin and Growth of Physical Sciences* (Baltimore: Penguin Books, 1964); David Knight, ed., *Classical Scientific Papers: Chemistry* (New York: American Elsevier, 1968); David Knight, ed., *Classical Scientific Papers: Chemistry, Second Series, On the Nature and Arrangement of the Chemical Elements* (London: Mills and Boon, 1970); Mary Jo Nye, ed., *The Question of the Atom: From the Karlsruhe Congress to the First Solvay Conference, 1860–1911, A Compilation of Primary Sources* (Los Angeles: Tomash, 1984); and Stephen John

Wright, ed., *Classical Scientific Papers: Physics* (New York: American Elsevier, 1965).

Among the most readable histories of modern chemistry and physics are O. Theodor Benfey, *From Vital Force to Structural Formulas* (1964; rpt., Philadelphia: Beckman Center for the History of Chemistry, 1992); William H. Brock, *The Norton History of Chemistry* (New York: Norton, 1994); Aaron J. Ihde, *The Development of Modern Chemistry* (New York: Harper and Row, 1964); John Hudson, *The History of Chemistry* (New York: Chapman and Hall, 1992); P. M. Harman, *Energy, Force and Matter: The Conceptual Development of Nineteenth-Century Physics* (Cambridge: Cambridge University Press, 1982); Emilio Segrè, *From Falling Bodies to Radio Waves: Classical Physicists and Their Discoveries* (San Francisco: Freeman, 1984) and *From X-Rays to Quarks: Modern Physicists and their Discoveries* (San Francisco: Freeman, 1980).

A pair of volumes in French is highly readable: Bernadette Bensaude-Vincent and Isabelle Stengers, *Histoire de la chimie* (Paris: Editions de la Découverte, 1993), which will be published in English translation by Harvard University Press, and Michel Biezunski, *Histoire de la physique moderne* (Paris: Editions de la Découverte, 1993).

Lewis Pyenson, *Civilizing Mission: Exact Sciences and French Overseas Expansion, 1830–1940* (Baltimore: Johns Hopkins University Press, 1993) is noted in the Introduction. On "Big Science" see Alvin M. Weinberg, "Impacts of Large-Scale Science," *Science* 134 (1961): 161–164; and Derek J. de Solla Price, *Little Science, Big Science* (New York: Oxford University Press, 1963) and Price, "The Exponential Curve of Science," *Discovery* 17 (1956): 240–243. Recent historical work on Big Science includes *Science after '40*, a special issue of *Osiris*, 2d series, (1992) edited by Arnold Thackray, especially, James H. Capshew and Karen A. Rader, "Big Science: Price to the Present," pp. 3–25, and Peter Galison and Bruce Hevly, eds. *Big Science: The Growth of Large-Scale Research* (Palo Alto, Calif.: Stanford University Press, 1992). Also Daniel J. Kevles, *The Physicists: The History of a Scientific Community in Modern America*, 2nd ed. (Cambridge, Mass.: Harvard University Press, 1995).

Chapter 1

For a comparison of national trends in scientific organization and institutions in the nineteenth century, see Paul Forman, John Heilbron, and Spencer Weart, eds., *Physics ca. 1900*, vol. 5, *Historical Studies in the Physical Sciences* (Princeton: Princeton University Press, 1975), and the classical volumes by John Merz, *A History of European Thought in the Nineteenth Century*, vols. 1 and 2, *Scientific Thought*, rpt. (New York: Dover, 1965).

On the development of disciplines in chemistry and physics, including laboratories, see Graeme Gooday, "Precision Measurement and the Genesis of Physics Teaching Laboratories in Victorian Britain," *British Journal for the History of Science* 23 (1990): 23–51; Ihde's *Development of Modern Chemistry* (1964), mentioned above; Frank A. J. L. James, ed., *The Development of the Laboratory: Essays on the Place of Experiment in Industrial Civilization* (New York: American Institute of Physics, 1989); Keith A. Nier, "The Emergence of Physics in Nineteenth-Century Britain as a Socially Organized Category of Knowledge: Preliminary Studies," Harvard University, Ph.D. diss., 1975; Mary Jo Nye, *From Chemical Philosophy to Theoretical Chemistry: Dynamics of Matter and Dynamics of Disciplines, 1800–1950* (Berkeley: University of California Press, 1993); Kathryn M. Olesko, *Physics as a Calling: Discipline and Practice in the Königsberg Seminar for Physics* (Ithaca, N.Y.: Cornell University Press, 1991); J. R. Partington, *A Short History of Chemistry* (London: Macmillan, 1957); Colin A. Russell, *The Structure of Chemistry* (Milton Keynes: Open University, 1976); and Anthony S. Travis, Willem J. Hornix, and Robert Bud, eds., *Organic Chemistry and High Technology, 1850–1950*, vol. 25 of the *British Journal for the History of Science* (1992).

Matthew Arnold gave a firsthand report on *Schools and Universities on the Continent* (London: Macmillan, 1868); see also B. N. Clark, "The Influence of the Continent upon the Development of Higher Education and Research in Chemistry in Great Britain during the Latter Half of the Nineteenth Century," University of Manchester Ph.D. diss., 1979; Robert Fox and George Weisz, eds., *The Organization of Science and Technology in France 1808–1914* (Cambridge: Cambridge University Press, 1980); Christa Jungnickel and Russell McCormmach, *Intellectual Mastery of Nature: Theoretical Physics from Ohm to Einstein*, 2 vols. (Chicago: University of Chicago Press, 1986); and Mary Jo Nye, *Science in the Provinces: Scientific Communities and Provincial Leadership in France, 1860–1930* (Berkeley: University of California Press, 1986).

On Great Britain, sources include Robert Bud and Gerrylynn K. Roberts, *Science versus Practice: Chemistry in Victorian Britain* (Manchester: Manchester University Press, 1984); David Gooding and Frank A. J. L. James, eds., *Faraday Rediscovered: Essays on the Life and Work of Michael Faraday, 1791–1867* (New York: American Institute of Physics, 1989); P. M. Harman, ed., *Wranglers and Physicists: Studies on Cambridge Mathematical Physics in the Nineteenth Century* (Manchester: Manchester University Press, 1985); and H. Bence Jones, *The Royal Institution: Its Founders and Its First Professors* (London: Longmans, Green, 1871).

Ludwig Boltzmann made firsthand observations of California in "On the Trip of a German Professor into El Dorado," in John Blackmore, ed., *Ludwig Boltzmann: His Later Life and Philosophy, 1900–1906* (Dordrecht:

Kluwer, 1995), pp. 171–197. On Americans studying in Germany, see Erwin N. Hiebert, "Nernst and Electrochemistry," in George Dubpernell and J. H. Westbrook, eds., *Selected Topics in the History of Electrochemistry* (Princeton: The Electrochemical Society, 1978), pp. 180–200. On chemistry and physics in the United States, see Alexandra Oleson and John Voss, eds., *The Organization of Knowledge in Modern America, 1860–1920* (Baltimore: Johns Hopkins University Press, 1979), and John W. Servos, *Physical Chemistry from Ostwald to Pauling: The Making of a Science in America* (Princeton: Princeton University Press, 1990). Daniel J. Kevles's *The Physicists: The History of a Scientific Commity in Modern America* (1971; Cambridge, Mass.: Harvard University Press, 1995) is unsurpassed.

The best history of women in science is Margaret W. Rossiter, *Women Scientists in America: Struggles and Strategies to 1940* (Baltimore: Johns Hopkins University Press, 1982) and *Women Scientists in America: Before Affirmative Action, 1940–1972* (Baltimore: Johns Hopkins University Press, 1995). For other studies on women in science, see Pnina Abir-Am, Helena Pycior, and Nancy Slack, eds., *Collaborative Couples in Biological, Physical and Social Sciences: Comparative Studies of Creativity and Intimacy* (New Brunswick, N.J.: Rutgers University Press, 1996). Marilyn B. Ogilvie's *Women in Science* is mentioned in the introduction. Gillian Sutherland's essay "Emily Davies, the Sidgwicks and the Education of Women in Cambridge," in Richard Mason, ed., *Cambridge Minds* (Cambridge: Cambridge University Press, 1994), is informative about efforts at educating women scientists in the old British universities.

Among sources for beginning the study of African-American chemists and physicists before 1940 are Vivian Ovelton Sammons's *Blacks in Science and Medicine* (1990), and Julius H. Taylor, ed., *The Negro in Science* (Baltimore: Morgan State College Press, 1955). Also see Kenneth R. Manning, "Race, Gender, and Science," in Henry Steffens, ed., *Topical Essays for Teachers* (Seattle, Washington: The History of Science Society, 1995), pp. 5–34.

Thomas Sprat wrote the first *History of the Royal Society* in 1667 (London: Routledge, Kegan and Paul, 1959), edited by Jackson I. Cope and Harold W. Jones. Other histories of scientific academies include the splendid book by Maurice Crosland, *Science under Control: The French Academy of Sciences, 1795–1914* (Cambridge: Cambridge University Press, 1992); Marie Boas Hall, *All Scientists Now: The Royal Society in the Nineteenth Century* (Cambridge: Cambridge University Press, 1984); and Armin Hermann, "Physiker und Physik—anno. 1845. 120 Jahre Physikalische Gesellschaft in Deutschland," *Physikalische Blätter* 21 (1969): 399–405. Two excellent sources on science publishing are William H. Brock and A. J. Meadows, *The Lamp of Learning: Taylor and Francis and the Development of Science Publishing* (London: Taylor and Francis, 1984), and A. J. Meadows, ed., *The Development of Science Publishing in Europe* (Amsterdam: Elsevier, 1980).

The best study of the origins of the Nobel Prizes is Elisabeth Crawford, *The Beginnings of the Nobel Institution: The Science Prizes 1901–1915* (Cambridge: Cambridge University Press, 1984). Equally valuable is her *Nationalism and Internationalism in Science, 1880–1939: Four Studies of the Nobel Population* (Cambridge: Cambridge University Press, 1992). I have also drawn on a summary of a paper by Paul R. Jones, "The Training in Germany of English-Speaking Chemists in the Nineteenth-Century and Its Profound Influence in Britain and America," which was presented at First International Summer Institute in the German Democratic Republic, Philosophy and History of Science, Leipzig, 29 June 1988.

Chapter 2

Ida Freund was a chemist who taught at Newnham College, which was established for women in Cambridge, England, in 1875. Her history of chemistry, *The Study of Chemical Composition: An Account of its Method and Historical Development* (1904; rpt., New York: Dover, 1968), remains a valuable one. As is the case in Freund's history, many excerpts from nineteenth-century chemists' papers and books are found in O. Theodor Benfey's *From Vital Force to Structural Formulas* (1992), mentioned among sources for the Introduction. The source for Newton's Query 31 is Isaac Newton, *Opticks*, 4th ed. (1730; rpt. New York: McGraw-Hill, 1931).

Other invaluable resources for this chapter that have already been mentioned are Ihde, *Development of Modern Chemistry* (1964); Partington, *Short History of Chemistry* (1957); and Russell, *The Structure of Chemistry* (1976). J. R. Partington's *A History of Chemistry*, vol. 4 (London: Macmillan, 1964), remains one of the most important and detailed histories of chemistry for the nineteenth century.

On Dalton, and on the chemical atom, see first of all Alan J. Rocke, *Chemical Atomism in the Nineteenth Century: From Dalton to Cannizzaro* (Columbus, Ohio: Ohio State University Press, 1984), as well as David M. Knight, *Atoms and Elements: A Study of Matter in England in the Nineteenth Century* (London: Hutchinson, 1967); Mary Jo Nye, "The Nineteenth-Century Atomic Debates and the Dilemma of an 'Indifferent Hypothesis,'" *Studies in the History and Philosophy of Science* 7 (1976): 245–268; and Arnold W. Thackray, "The Emergence of Dalton's Atomic Theory," *British Journal for the History of Science* 3 (1966): 1–23.

Among the general sources for the history of physics and physical sciences, as used in this chapter, are ones mentioned earlier: Brush, *The History of Modern Science* (1988); Harman, *Energy, Force, and Matter* (1982); Jungnickel and McCormmach, *Intellectual Mastery of Nature* (1986); and

Segrè, *From Falling Bodies to Radio Waves* (1984). An excellent resource originally written for physics students is Gerald Holton, with revisions by Stephen G. Brush, *Introduction to Concepts and Theories in Physical Science*, 2d ed. (Princeton: Princeton University Press, 1985).

On the Laplacian tradition of Newtonian natural philosophy, see Maurice Crosland, *Science under Control: The French Academy of Sciences, 1795–1914* (Cambridge: Cambridge University Press, 1992); Robert Fox, "The Rise and Fall of Laplacian Physics," in *Historical Studies in the Physical Sciences*, vol. 4 (Princeton: Princeton University Press, 1974), pp. 89–136. On quarks, see Michael Riordan, "The Discovery of Quarks," *Science* 256 (29 May 1992): 1287–1293.

Chapter 3

Also of value for the history of theories of electromagnetism and the ether are the previously mentioned Gillispie, *The Edge of Objectivity* (1960); Harman, *Energy, Force, and Matter* (1982): Harman, ed., *Wranglers and Physicists* (1985); Jungnickel and McCormmach, *Intellectual Mastery of Nature* (1986); and Segrè, *From Falling Bodies to Radio Waves* (1984). For some primary sources, see Nye, ed., *The Question of the Atom* (1984).

The *Bulletin for the History of Chemistry* 11 (Winter 1991) consists of an excellent group of articles collectively entitled *Michael Faraday—Chemist and Popular Lecturer.* Other sources used on Faraday include Geoffrey Cantor, *Michael Faraday: Sandemanian and Scientist* (London: Macmillan, 1991); David Gooding and Frank A. J. L. James, eds., *Faraday Rediscovered: Essays on the Life and Work of Michael Faraday, 1791–1867* (New York: American Institute of Physics, 1989), and David Gooding, Trevor Pinch, and Simon Schaffer, eds., *The Uses of Experiment: Studies in the Natural Sciences* (Cambridge: Cambridge University Press, 1989), especially Gooding's article "'Magnetic Curves' and the Magnetic Field," pp. 182–223, and James's article "'The Optical Mode of Investigation': Light and Matter in Faraday's Natural Philosophy," pp. 137–162. For a strongly interpretative biography, see L. Pearce Williams, *Michael Faraday* (London: Chapman and Hall, 1965). Faraday's *Experimental Researches in Electricity*, 3 vols. (London: Quaritch, 1839–1855), are a primary resource.

For Maxwell and electromagnetism after Maxwell, including the work of Heinrich Hertz, Jed Z. Buchwald's work is among the most important, including "Modifying the Continuum: Methods of Maxwellian Electrodynamics," in Harman, ed., *Wranglers and Physicists* (1985), pp. 225–241; *From Maxwell to Microphysics: Aspects of Electromagnetic Theory in the Last Quarter of the Nineteenth Century* (Chicago: University of Chicago Press, 1985); "Electrodynamics in Context: Object States, Laboratory Practice, and Anti-Romanticism," in David Cahan, ed., *Hermann von Helmholtz and the Foundations of Nineteenth-Century Science* (Berkeley: University of California Press, 1993), pp. 334–373; and *The Creation of Scientific Effects: Heinrich Hertz and Electric Waves* (Chicago: University of Chicago Press, 1994). Robert S. Cohen's introduction is valuable in Heinrich Hertz, *The Principles of Mechanics Presented in a New Form,* translated D. E. Jones and J. T. Walley (New York: Dover, 1956).

Also on Maxwell, see C. W. F. Everitt, *James Clerk Maxwell: Physicist and Natural Philosopher* (New York: Charles Scribner's Sons, 1975), which is a version of Everitt's article on Maxwell in the *Dictionary of Scientific Biography.* On the Cavendish Laboratory, see Jeffrey Hughes, "'Brains in their Fingertips': Physics at the Cavendish," in Richard Mason, ed., *Cambridge Minds* (Cambridge: Cambridge University Press, 1994), pp. 160–176.

For some of Maxwell's most significant writings, see W. D. Niven, ed., *The Scientific Papers of James Clerk Maxwell,* 2 vols. (Cambridge: Cambridge University Press, 1890), rpt., including, in vol. 1, "On Faraday's Lines of Force" (1856), pp. 155–229; "On Physical Lines of Force," Pt. I and Pt. II (1861), pp. 451–488, and Pt. III (1862), pp. 489–513; and "A Dynamical Theory of the Electromagnetic Field" (1864), pp. 526–597. See also, Maxwell's *Matter and Motion* (1877; rpt., New York: Dover, n.d.) and *An Elementary Treatise on Electricity,* edited by William Garnett (Oxford: Clarendon, 1881).

On William Thomson (Lord Kelvin), the fundamental biographical study is Crosbie Smith and M. Norton Wise, *Energy and Empire: A Biographical Study of Lord Kelvin* (Cambridge: Cambridge University Press, 1989). Quotation is made from Thomson's "On Vortex Atoms," *Philosophical Magazine* [4], 34 (1867): 15–24. See also Ole Knudsen, "Mathematics and Physical Reality in William Thomson's Electromagnetic Theory," in Harman, ed., *Wranglers and Physicists* (1985), pp. 149–179.

On Helmholtz, there is an extensive and valuable collection of studies in Cahan, ed., *Helmholtz* (1993), mentioned above. Helmholtz's "On the Modern Development of Faraday's Conception of Electricity," in C. S. Gibson and A. J. Greenaway, eds., *Faraday Lectures 1869–1928* (London: The Chemical Society, Burlington House, 1928), pp. 132–159, remains eminently readable.

A valuable resource for the British electromagnetic tradition is Bruce J. Hunt, *The Maxwellians* (Ithaca, N. Y.: Cornell University Press, 1991); on Lorentz's electrodynamics, see Russell McCormmach, "H. A. Lorentz and the Electromagnetic View of Nature," *Isis* 61 (1970): 459–497; and on experimental tests for the ether, see Gerald Holton, *Thematic Origins of Scientific Thought: Kepler to Einstein* (Cambridge, Mass.: Harvard University Press, 1973). A fascinating article on the many uses of the ether is Helge Kragh, "The Aether in Late Nineteenth-Century Chemistry," *Ambix* 36 (1989): 49–65. See also the original papers by William Crookes,

"On Radiant Matter," *American Journal of Science* 118 (1879), 241–262, and J. Norman Lockyer, "Atoms and Molecules Spectroscopically Considered," in *Studies in Spectrum Analysis*, 2d ed. (London: Kegan Paul, 1878), pp. 113–144.

On scientific styles, excerpts are from J. B. Birks, ed., *Rutherford at Manchester* (London: Heywood, 1962), and Pierre Duhem, *The Aim and Structure of Physical Theory,* translated by Philip Wiener (Princeton: Princeton University Press, 1954).

CHAPTER 4

As in earlier chapters, some of the more general sources important for this chapter are Harman, *Energy, Force, and Matter* (1982); Hudson, *History of Chemistry* (1992); Jungnickel and McCormmach, *Intellectual Mastery of Nature* (1986); McKenzie, *Major Achievements of Science* (1988); Nye, ed., *Question of the Atom* (1984); Segrè, *From Falling Bodies to Radio Waves* (1984); and Segrè, *From X-Rays to Quarks* (1980).

On the complexity of the problem of the "discovery" of conservation of energy, see Thomas S. Kuhn, "Energy Conservation as an Example of Simultaneous Discovery," in Marshall Clagett, ed., *Critical Problems in the History of Science,* (Madison, Wis.: University of Wisconsin Press, 1959), pp. 321–356, as well as Yehuda Elkana, *The Discovery of the Conservation of Energy* (London: Hutchinson, 1974). On Joule, see Donald S. L. Cardwell, "The Origins and Consequences of Certain of J. P. Joule's Scientific Ideas," in Rutherford Aries, H. Ted David, and Roger H. Stuewer, eds., *Springs of Scientific Creativity: Essays on Founders of Modern Science* (Minneapolis: University of Minnesota Press, 1983), pp. 44–70; J. G. Crowther, "James Prescott Joule, 1818–1889," in Crowther's *Men of Science* (1936), pp. 127–197; and Robert Kargon, *Science in Victorian Manchester: Enterprise and Expertise* (Baltimore: Johns Hopkins University Press, 1977).

On Helmholtz, see Fabbio Bevilacqua, "Helmholtz's 'Ueber die Erhaltung der Kraft,'" in Cahan, ed., *Helmholtz* (1993); on Thomson (Lord Kelvin), Smith and Wise, *Energy and Empire* (1989); and Everitt, *James Clerk Maxwell* (1975), as well as Silvanus P. Thomson, *The Life of William Thomson, Baron Kelvin of Largs,* 2 vols. (London, Macmillan, 1910). Among sources for Thomson's papers are his *Mathematical and Physical Papers,* 6 vols. (Cambridge: Cambridge University Press, 1882–1911) and *Baltimore Lectures on Molecular Dynamics and the Wave Theory of Light* (Cambridge: Cambridge University Press, 1904), based on his lectures at the Johns Hopkins University in 1884.

On thermochemistry and chemical thermodynamics, see Diana K. Barkan, "The Witches' Sabbath: The First International Solvay Congress in Physics," *Science in Context* 6 (1993): 59–82; Erwin N. Hiebert, "Walther Nernst and the Application of Physics to Chemistry," in Aries et al., *Springs of Scientific Creativity* (1983), pp. 203–231; Mi Gyung Kim, "Practice and Representation: Investigative Programs of Chemical Affinity in the Nineteenth Century," University of California at Los Angeles, Ph.D. diss., 1990; Helge Kragh, "Julius Thomsen and Classical Thermochemistry," *British Journal for the History of Science* 17 (1984): 255–272; Helge Kragh, "Between Physics and Chemistry: Helmholtz's Route to a Theory of Chemical Thermodynamics," in Cahan, ed., *Helmholtz* (1993), pp. 403–431; and V. V. Raman, "The Permeation of Thermodynamics into Nineteenth-Century Chemistry," *Indian Journal of History of Science* 10 (1975): 16–37.

For primary sources, see Claude-Louis Berthollet, *Researches into the Law of Chemical Affinity*, translated by M. Farrell (Baltimore: Philip Necklin, 1809); Pierre Duhem, *Traité élémentaire de mécanique chimique* (Paris: Hermann, 1897); Wilhelm Ostwald, "Elements and Compounds," in Gibson and Greenaway, eds., *Faraday Lectures* (1928), pp. 185–201, and J. J. Thomson, *Application of Dynamics to Physics and Chemistry* (London: Macmillan, 1888).

Stephen G. Brush has written on the "Foundations of Statistical Mechanics, 1845–1915," *Archives for History of Exact Sciences* 4 (1967): 145–183; as have Lorraine Daston, *Classical Probability in the Enlightenment* (Princeton: Princeton University Press, 1988), and Theodore M. Porter, *The Rise of Statistical Thinking* (Princeton: Princeton University Press, 1986). See also Martin J. Klein, "The Scientific Style of Josiah Willard Gibbs," in Aries et al., *Springs of Scientific Creativity* (1983), pp. 142–162, and original articles by Robert Clausius, "Le second principe fondamental de la théorie mécanique de la chaleur," *Revue des cours scientifiques* 5 (8 February 1868): 153–159, translated by P. Delestrée; James Clerk Maxwell, "A Discourse on Molecules," *Philosophical Magazine* 46 (1873): 453–469; and James Clerk Mazwell, "On the Dynamical Evidence of the Molecular Constitution of Bodies," *Journal of the Chemical Society (London)* 28 (1875): 493–508. On kinetics and chemistry as viewed by an eminent physical chemist, see Keith J. Laidler, "Chemical Kinetics and the Origins of Physical Chemistry," *Archive for History of Exact Sciences* 32 (1985): 43–75; and Keith J. Laidler, *The World of Physical Chemisty* (Oxford: Oxford University Press, 1993).

John Heilbron's *The Dilemmas of an Upright Man: Max Planck as a Spokesman for German Science* (Berkeley: University of California Press, 1986) is an unparalleled short biography of Planck, and Thomas S. Kuhn's *Black-Body Theory and the Quantum Discontinuity, 1894–1912* (Oxford: Clarendon, 1978) is the definitive analysis of the early history of quantum theory. On German institutions where this work was carried out, see David Cahan, *An Institute for an Empire: The Physikalisch-Technische Reichsanstalt 1871–1918* (Cambridge: Cambridge University Press, 1989), and Jeffrey Allan Johnson, *The Kaiser's Chemists: Science and Modernization in Imperial Germany* (Chapel Hill, N.C.: University of North Carolina Press, 1990).

Einstein's papers on Brownian motion are found in *Investigations on the Theory of the Brownian Movement*, edited by R. Fürth and translated by A. D. Cowper (New York: Dover, 1956). The significance of Jean Perrin's work is treated in Mary Jo Nye, *Molecular Reality: A Perspective on the Scientific Work of Jean Perrin* (London: Macdonald, and New York: Elsevier, 1972). See also Nye's "The Nineteenth-Century Atomic Debates and the Dilemma of an 'Indifferent Hypothesis,'" *Studies in the History and Philosophy of Science* 7 (1976): 254–268.

On criticisms of van't Hoff, see Alan J. Rocke, *The Quiet Revolution: Hermann Kolbe and the Science of Organic Chemistry* (Berkeley: University of California Press, 1993). On anxiety and crisis, see Phillip Frank, *Modern Science and Its Philosophy* (Cambridge, Mass.: Harvard University Press, 1950); Mary Jo Nye, "Gustave Le Bon's Black Light: A Study of Physics and Philosophy in France at the Turn of the Century," *Historical Studies in the Physical Sciences* 4 (1974): 163–195; and, more powerfully, H. G. Wells, *The Time Machine: An Invention* (1895; many editions).

Chapter 5

These sources also have been mentioned for earlier chapters: Benfey, *From Vital Force to Structural Formula* (1992); Brock, *The Norton History of Chemistry* (1993); Freund, *The Study of Chemical Composition* (1968); Hudson, *The History of Chemistry* (1992); Ihde, *The Development of Modern Chemistry* (1964); Johnson, *The Kaiser's Chemists* (1990); Nye, *Science in the Provinces* (1986) and *From Chemical Philosophy to Theoretical Chemistry* (1993); Partington, *A History of Chemistry*, vol. 4 (1964); Russell, *The Structure of Chemistry* (1976); and Servos, *Physical Chemistry from Ostwald to Pauling* (1990).

Alan J. Rocke's *The Quiet Revolution* (1993), while focusing on Hermann Kolbe, is an indispensable history of German chemistry in the nineteenth century. See, too, Jeffrey A. Johnson, "Hierarchy and Creativity in Chemistry, 1871–1914," *Osiris*, 2d series, 5 (1989): 214–240 (part of the special issue on *Science in Germany*, noted below). A strong, but by no means exclusively French perspective is found in Bensaude-Vincent and Stengers, *Histoire de la chimie* (1993), mentioned above. A valuable interpretative history is David Knight, *The Transcendental Part of Chemistry* (Folkestone, Kent, 1978).

On aspects of chemistry and the natural history tradition, sources include Toby Appell, *The Cuvier-Geoffroy Debate: French Biology in the Decades before Darwin* (Oxford: Oxford University Press, 1987) and Frederic L. Holmes, *Eighteenth-Century Chemistry as an Investigative Enterprise* (Berkeley: Office for History of Science and Technology at University of California at Berkeley, 1989). On the evolution of the type and structural theories, see Mary Ellen Bowden and Theodor Benfey, *Robert Burns Woodward and the Art of Organic Synthesis* (Philadelphia: Beckman Center for the History of Chemistry, 1992); Michael N. Keas, "The Structure and Philosophy of Group Research: August Wilhelm Hofmann's Research Program in London (1845–1865)," University of Oklahoma, Ph.D. diss., 1992; Mi Gyung Kim, "Practice and Representation: Investigative Programs of Chemical Affinity in the Nineteenth Century," University of California at Los Angeles, Ph.D. diss., 1990; Mi Gyung Kim, "The Layers of Chemical Language, I: Constitution of Bodies *v.* Structure of Matter," *History of Science* 30 (1992): 60–96; and D. Stanley Tarbell, "Organic Chemistry: The Past 100 Years," *Chemical and Engineering News* 54 (6 April 1976): 110–123.

Historical and sociological literature on chemical research schools is important and informative both about the chemical discipline and about methods and theories in chemistry. See first of all the work of distinguished biochemist Joseph S. Fruton, "Contrasts in Scientific Style. Emil Fischer and Franz Hofmeister: Their Research Groups and Their Theory of Protein Structure," *Proceedings of the American Philosophical Society* 129 (1985): 313–370, and "The Liebig Research Group—A Reappraisal," *Proceedings of the American Philosophical Society* 132 (1988): 1–66.

An important essay on research schools is J. B. Morrell, "The Chemist Breeders: The Research Schools of Liebig and Thomson," *Ambix* 19 (1972): 1–46, followed by Gerald L. Geison, "Scientific Change, Emerging Specialties, and Research Schools," *History of Science* 19 (1981): 20–40. Gerald L. Geison and Frederic L. Holmes have edited the special issue of *Osiris* (2d series, 8 [1993]) entitled *Research Schools: Historical Reappraisals,* which includes J. B. Morrell, "W. H. Perkin, Jr., at Manchester and Oxford: From Irwell to Isis," 104–126, and John W. Servos, "Research Schools and Their Histories," 3–15. See also Frederic L. Holmes, "The Complementarity of Teaching and Research in Liebig's Laboratory," pp. 121–164 of the special issue of *Osiris* (2d series, 5 [1989]) edited by Kathryn M. Olesko and entitled *Science in Germany: The Intersection of Institutional and Intellectual Issues,* as well as L. J. Klosterman, "A Research School of Chemistry in the Nineteenth Century: Jean-Baptiste Dumas and His Research Students," *Annals of Science* 43 (1985): 1–80.

On Kekulé, see Richard Anschütz's two-volume biography *August Kekulé* (Berlin: Verlag Chemie, 1929), as well as Kekulé's "The Scientific

Aims and Achievements of Chemistry," *Nature* 18 (1878): 210–213. Other primary sources include Henry Armstrong, "Presidential Address," Chemistry Section, *BAAS Reports, Winnipeg (1909)* (1910): 420–454; Marcellin Berthelot, *Leçons sur les méthodes générales de synthèse organique* (Paris: Gauthier-Villars, 1864); Jean-Baptiste-André Dumas, "Mémoire sur la constitution de quelques corps organiques et sur la théorie de substitutions," *Comptes rendus hebdomadaires de l'Académie des Sciences* (1839): 609–622, and "Mémoire sur la loi des substitutions et la théorie des types," *Comptes rendus* (1840): 149–178; Charles Friedel, *Cours de chimie organique, professé à la Faculté des Sciences* (Paris, 1887).

On Pasteur, stereochemistry, and molecular architecture, sources are René J. Dubos, *Louis Pasteur: Free Lance of Science* (Boston: Little, Brown and Co., 1950); Gerald L. Geison and James Secord, "Pasteur and the Process of Discovery: The Case of Optical Isomerism," *Isis* 79 (1988): 6–36; Jean Jacques, *The Molecule and Its Double,* translated by Lee Scanlon (New York: McGraw-Hill, 1993); Bruno Latour, *Pasteur, une science, un style, un siècle* (Paris: Perrin, 1994); and Craig Zwerling, "The Emergence of the École Normale Supérieure as a Centre of Scientific Education in the Nineteenth Century," in Fox and Weisz, eds., *The Organization of Science and Technology in France* (1980), pp. 31–60. P. Vallery-Radot has edited the *Oeuvres de Pasteur,* including vol. 1, *Dissymétrie moléculaire* (Paris: Masson, 1922). More recent is Gerald L. Geison, *The Private Science of Louis Pasteur* (Princeton: Princeton University Press, 1995).

On the history of proteins and bigger molecules, see Yasu Furukawa, "Hermann Staudinger and the Emergence of the Macromolecular Concept," *Historia Scientiarum,* 22 (1982): 1–18. A revision of Furukawa's study (based in his University of Oklahoma Ph.D. dissertation of 1983) on Hermann Staudinger, W. H. Carothers, and macromolecular chemistry is being published by the University of Pennsylvania Press. A personal view is given by Herman F. Mark, "Polymer Chemistry: The Past 100 Years," *Chemical and Engineering News* 54 (6 April 1976): 176–189, and in an "Interview with Herman F. Mark," *Journal of Chemical Education* 56 (1979), 83–86.

The history of biochemistry figures in Frederic L. Holmes, *Between Biology and Medicine: The Formation of Intermediary Metabolism* (Berkeley: Office for History of Science and Technology at University of California at Berkeley, 1992), and in Robert E. Kohler, *Partners in Science: Foundations and Natural Scientists 1900–1945* (Chicago: University of Chicago Press, 1991). See, too, physical chemist Michael Polanyi's "My Time with X-Rays and Crystals," in P. P. Ewald, ed., *Fifty Years of X-Ray Diffraction* (Utrecht: International Union of Crystallography, 1962), pp. 629–636, as well as Alexander Todd's *A Time to Remember: The Autobiography of a Chemist* (Cambridge: Cambridge University Press, 1983).

On Linus Pauling and research at Caltech, see Judith R. Goodstein, *Millikan's School: A History of the California Institute of Technology* (New York: Norton, 1991), as well as Thomas Hager, *Force of Nature: The Life of Linus Pauling* (New York: Simon and Schuster, 1995). Pnina Abir-Am is studying the scientific work of Dorothy C. Hodgkin, as noted in P. Abir-Am, "Women in Research Schools: Approaching an Analytical Lacuna in the History of Chemistry and Allied Sciences," in Seymour H. Mauskopf, ed., *Chemical Sciences in the Modern World* (Philadelphia: University of Pennsylvania Press, 1993), pp. 375–391.

I have always found useful both for its chemistry and for its historical aperçus Louis F. Fieser and Mary Fieser, *Introduction to Organic Chemistry* (Boston: D. C. Heath, 1957).

Chapter 6

General sources for this chapter again include Holton, *Introduction to Concepts and Theories in Physical Science* (1972); Ihde, *The Development of Modern Chemistry* (1964); McKenzie, *The Major Achievements of Science* (1960); Nye, ed., *The Question of the Atom* (1984) and *From Chemical Philosophy to Theoretical Chemistry* (1993); Segrè, *From X-Rays to Quarks* (1980); and Servos, *Physical Chemistry from Ostwald to Pauling* (1990).

Books focused on quantum mechanics include the personal account of George Gamow, *Thirty Years That Shook Physics: The Story of Quantum Theory* (1966; rpt., New York: Dover, 1985); Max Jammer, *The Conceptual Development of Quantum Mechanics* (New York: McGraw-Hill, 1966); and, as an edition of original papers, Dirk ter Haar, ed., *The Old Quantum Theory* (Oxford: Pergamon Press, 1967). The physicist Steven Weinberg's *The Discovery of Subatomic Particles* (New York: W. H. Freeman, 1993) is useful.

On electron theory, quantum mechanics, and theories of chemical bonding, see Colin A. Russell, *A History of Valency* (New York: Humanities Press, 1971), and Anthony Stranges, *Electrons and Valence* (College Station, Tex.: Texas A&M University Press, 1982).A textbook in physical chemistry of considerable influence was Farrington Daniels and Robert A. Alberty, *Physical Chemistry*, 2d ed. (New York: Wiley, 1963).

Laylin K. James has edited a very useful volume of biographical essays, *Nobel Laureates in Chemistry: 1901–1992* (Philadelphia: American Chemical Society and Chemical Heritage Foundation, 1993). Included is a fine, short study of "Marie Curie" by Bernadette Bensaude-Vincent (pp. 75–82). Robert Reid's *Marie Curie* (New York: New American Library, 1974) is one of the best popular biographies of Curie. A more recent and very fine biography is Susan Quinn, *Marie Curie: A Life* (New York: Simon and Schuster, 1995).

A fine account of early theories of X rays is Bruce R. Wheaton, *The Tiger and the Shark: Empirical Roots of Wave-Particle Dualism* (Cambridge: Cambridge University Press, 1983). Among the many sources on early studies of X rays, electrons, ions, and radioactivity are J. B. Birks, ed., *Rutherford at Manchester* (London: Heywood, 1962); Jun Fudano, "Early X-Ray Research at Physical Laboratories in the United States of America *circa* 1900: A Reappraisal of American Physics," University of Oklahoma, Ph.D. diss., 1990; Peter Galison and Alexi Assmus, "Artificial Clouds, Real Particles," in Gooding, Pinch, and Schaffer, eds., *The Uses of Experiment* (1989); John L. Heilbron, *H. G. J. Moseley: The Life and Letters of an English Physicist, 1887–1915* (Berkeley: University of California Press, 1974); Nye, *Molecular Reality* (1972), mentioned above; Abraham Pais, "Radioactivity's Two Early Puzzles," *Reviews of Modern Physics* 49 (1977): 925–938; and Alexander Wood, *The Cavendish Laboratory* (Cambridge: Cambridge University Press, 1946).

The spurious N rays and black light are analyzed in Irving M. Klotz, "The N-Ray Affair," *Scientific American* 242 (1980): 168–175; Mary Jo Nye, "Gustave Le Bon's Black Light: A Study in Physics and Philosophy in France at the Turn of the Century," in *Historical Studies in the Physical Sciences,* vol. 4 (Princeton: Princeton University Press, 1974), pp. 163–195; and Mary Jo Nye, "N-Rays: An Episode in the History and Psychology of Science," *Historical Studies in the Physical Sciences* 9 (1980): 125–156.

Original papers by Rutherford include the following: Ernest Rutherford and Frederick Soddy, "Radioactive Change," *Philosophical Magazine,* series 6, 5 (1903): 576–591 (reprinted in Nye, ed., *Question of the Atom,* pp. 605–624); Rutherford, "Present Problems of Radioactivity" (1904), in Katherine R. Sopka, ed., *Physics for a New Century: Papers Presented at the 1904 St. Louis Congress* (Los Angeles: Tomash, and New York: American Institute of Physics, 1986), pp. 133–262, and "The Scattering of *alpha* and *beta* Particles by Matter and the Structure of the Atom," *Philosophical Magazine,* series 6, 21 (1911): 669–688, also reprinted in *Question of the Atom.* Also relied on in the chapter is J. J. Thomson, *Electricity and Matter* (Westminster: Constable, 1904).

The physicist Abraham Pais is one of the recent biographers of Bohr; see his *Niels Bohr's Times* (Oxford: Oxford University Press, 1992). The best study of the origins of Bohr's atom remains John L. Heilbron and Thomas S. Kuhn, "The Genesis of the Bohr Atom," in *Historical Studies in the Physical Sciences,* vol. 1 (Princeton: Princeton University Press, 1969), pp. 211–290. John L. Heilbron, "The Rutherford-Bohr Atom," *American Journal of Physics* 49 (1981): 223–231, is also valuable.

Bohr's famous paper "On the Constitution of Atoms and Molecules" appeared in three consecutive issues of volume 26 of the *Philosophical Magazine* in 1913: part 1, "Binding of Electrons by Positive Nuclei," 1–25;

part 2, "Systems Containing Only a Single Nucleus," 476–502; and part 3, "Systems Containing Several Nuclei," 857–875.

On quantum theory among physicists in the 1920s, see Cassidy, *Uncertainty* (1992), mentioned above; Werner Heisenberg, *Encounters with Einstein and Other Essays on People, Places, and Particles* (Princeton: Princeton University Press, 1983); and Helge S. Kragh, *Dirac: A Scientific Biography* (Cambridge: Cambridge University Press, 1990). Original papers cited are P. A. M. Dirac, "Quantum Mechanics of Many-Electron Systems," *Proceedings of the Royal Society of London*, series A, 123 (1929): 714–733, and Ralph Fowler, "A Report on Homopolar Valency and Its Quantum-Mechanical Interpretation," in *Chemistry at the Centenary (1931) Meeting of the British Association for the Advancement of Science* (Cambridge: W. Heffer and Sons, 1932), pp. 226–246.

Anatole Abragam was briefly a student and later a colleague of Louis de Broglie. He wrote the critical and informative "Louis Victor Pierre Raymond de Broglie, 1892–1987," *Biographical Memoirs of Fellows of the Royal Society* 34 (1988): 22–41; another notice is by Bernard d'Espagnat, "Louis de Broglie (1892–1987)," *Nature* 327 (18 May 1987): 283. Georges Lochak has written a strongly partisan portrait, *Louis de Broglie* (Paris: Flammarion, 1992), which is complemented by the placing of de Broglie's work in the context of very contemporary physics by Peter R. Holland, *The Quantum Theory of Motion: An Account of the de Broglie-Bohm Causal Interpretation of Quantum Mechanics* (Cambridge: Cambridge University Press, 1993). De Broglie's thesis has been reprinted: *Recherches sur la théorie des quanta*, 3d ed. (1924; rpt., Paris: Fondation Louis de Broglie, 1992).

Original papers on electron theory and valence theory in chemistry are Irving Langmuir, "The Arrangement of Electrons in Atoms and Molecules," *Journal of the American Chemical Society* 41 (1919): 868–934, and Langmuir, "The Structure of Molecules," *BAAS Reports, Edinburgh (1921)* (1922): 468–469. Of considerable influence was G. N. Lewis, *Valence and the Structure of Atoms and Molecules* (Washington, D.C.: American Chemical Society, 1923). In organic chemistry, Christopher K. Ingold's "Principles of an Electronic Theory of Organic Reactions," *Chemical Reviews* 15 (1934): 225–274, and his *Structure and Mechanism in Organic Chemistry* (Ithaca, N.Y.: Cornell University Press, 1953) were widely read.

Americans led in the development of quantum chemistry. Accounts are given in Yuko Abe, "Pauling's Revolutionary Role in the Development of Quantum Chemistry," *Historia Scientiarium* 20 (1981): 107–124; Peter J. Hall, "The Pauli Exclusion Principle and the Foundations of Chemistry," *Synthèse* 16 (1986): 267–272; S. S. Schweber, "The Young John Clarke Slater and the Development of Quantum Chemistry," *Historical Studies in the Physical Sciences* 20 (1990): 339–406; Richard Severo, "Linus C. Pauling Dies at 93: Chemist and Voice for Peace," *The New York Times*, 21 August 1994, pp. 1, 16; Ana Isabel Simoes, "Converging Trajectories, Diverging Traditions: Chemical Bond, Valence, Quantum Mechanics and Chemistry, 1927–1937," University of Maryland, Ph.D. diss., 1993, and Kostas Gavroglu and Ana Isabel Simoes, "The Americans, the Germans, and the Beginnings of Quantum Chemistry: The Confluence of Diverging Traditions," *Historical Studies in the Physical and Biological Sciences* 25 (1994): 47–110. A personal account is given by Robert Mulliken, "Spectroscopy, Molecular Orbitals, and Chemical Bonding," in *Nobel Lectures. Chemistry. 1963–1970* (Amsterdam: Elsevier, 1972), pp. 131–160.

Pauling's scientific papers and books include "The Shared-Electron Chemical Bond," *Proceedings of the National Academy of Sciences* 14 (1928), 359–362: *The Nature of the Chemical Bond* (Ithaca, N.Y.: Cornell University Press, 1939); "The Nature of the Theory of Resonance," in Alexander Todd, ed., *Perspectives in Organic Chemistry* (New York: Intersciences Publishers, 1962), pp. 1–8; and, with E. Bright Wilson, Jr., *Introduction to Quantum Mechanics with Applications to Chemistry* (1935; rpt., New York: Dover, 1963).

Chapter 7

Among sources once again consulted are Cassidy, *Uncertainty* (1992); Gamow, *Thirty Years That Shook Physics* (1966); Holton, *Thematic Origins of Scientific Thought* (1988); Ihde, *Development of Modern Chemistry* (1964); Laylin, ed., *Nobel Laureates in Chemistry* (1993); Johnson, *The Kaiser's Chemists* (1990); Kevles, *The Physicists* (1995); Nye, *Science in the Provinces* (1986) and *From Chemical Philosophy to Theoretical Chemistry* (1993); Rossiter, *Women Scientists in America* (1982); Segrè, *From X-Rays to Quarks* (1980); and Sopka, ed., *Physics for a New Century* (1986). See also Marks, *Science and the Making of the Modern World* (1983), and Biezunski, *Histoire de la physique moderne* (1993).

Brigitte Schroeder-Gudehus persuasively analyzed the impact of World War I on the scientific community in *Les scientifiques et la paix. La communauté scientifique international au cour des années 20* (Montreal: Presses de l'Université de Montreal, 1978). The topic is also discussed by Daniel J. Kevles, "Into Hostile Camps: The Reorganization of International Science in World War I," *Isis* 62 (1971): 47–60.

Other accounts of scientists' involvement in World War I and the breakdown of internationalism through the mid-1920s are found in Crawford, *Nationalism and Internationalism in Science* (1992), and Heilbron, *The Dilemmas of an Upright Man* (1986), both mentioned above, as well as Daniel P. Jones, "American Chemists and the Geneva Protocol," *Isis* 71 (1980): 426–440, and W. A. E. McBryde, "Fritz Haber," in James, ed., *Nobel Laureates* (1993), pp. 114–123. Of interest, too, is the data on

Nobel Prize nominators and nominees in Elisabeth Crawford, J. L. Heilbron, and Rebecca Ullrich, *The Nobel Population, 1901–1937* (Berkeley: Office for History of Science and Technology at University of California at Berkeley, 1987).

Accounts of the origins of the Solvay conferences can be found in Maurice de Broglie, *Les Premiers Congrès de Physique Solvay et l'orientation de la physique depuis 1911* (Paris: Michel, 1951), and *Cinquantenaire de l'Institut Internationale de Chimie fondé par Ernest Solvay, 1913–1963* (Brussels: Weissenbruch, n.d.).

On Hermann Staudinger, see Yasu Furukawa, "Hermann Staudinger and the Emergence of the Macromolecular Concept" (1982), as well as his "Staudinger, Polymers, and Political Struggles," *Chemical Heritage* 11, no.1 (1993–1994): 4–6, and [in Japanese] *Kagakushi* [Journal of the Japanese Society for the History of Chemistry] 20 (1993): 1–19; also, Laylin K. James, "Hermann Staudinger," in James, ed., *Nobel Laureates* (1993), pp. 359–367.

Sources for the study of Einstein, his life, times, and work include Ronald W. Clark, *Einstein: The Life and Times* (New York: Avon, 1972); John Earman and Clark Glymour, "Relativity and Eclipses: The British Eclipse Expeditions of 1919 and their Predecessors," *Historical Studies in the Physical Sciences* 11 (1980): 49–85; Alan J. Friedman and Carol C. Donley, *Einstein as Myth and Muse* (Cambridge: Cambridge University Press, 1985); Gerald Holton, "Of Love, Physics and Other Passions: The Letters of Albert and Mileva," Parts 1 and 2, *Physics Today* 47 (August 1994): 23–29 and 47 (September 1994): 37–43; Banesh Hoffman, with Helen Dukas, *Albert Einstein: Creator and Rebel* (New York; National American Library, 1972); Abraham Pais, *'Subtle is the Lord . . . : The Science and the Life of Albert Einstein* (Oxford: Oxford University Press, 1982), and *Einstein Lived Here: Essays for the Layman* (Oxford: Clarendon, 1994): as well as Paul Arthur Schilpp, ed., *Albert Einstein: Philosopher-Scientist*, 2 vols. (New York: Harper, 1959).

For Einstein's own writings, the definitive edition is now in the process of publication: John Stachel, ed., *The Collected Papers of Albert Einstein* (Princeton: Princeton University Press, 1987–); I have also used Albert Einstein, Boris Podolsky, and Nathan Rosen, "Can Quantum-Mechanical Description of Physical Reality Be Considered Complete?" *Physical Review* 47 (1935): 777–780; and Albert Einstein and Leopold Infeld, *The Evolution of Physics: From Early Concepts to Relativity and Quanta* (New York: Simon and Schuster, 1938).

On physics and politics in the 1920 and 1930s, in addition to many of the sources cited above, I have drawn on J. G. Crowther, *Fifty Years with Science* (London: Barrie and Jenkins, 1970); Paul Forman, "Weimar Culture, Causality, and Quantum Theory, 1918–1927: Adaptation by German

Physicists and Mathematicians to a Hostile Intellectual Environment," *Historical Studies in the Physical Sciences* 3 (1971): 1–115; Mary Jo Nye, "Science and Socialism: The Case of Jean Perrin in the Third Republic," *French Historical Studies* 9 (Spring 1975): 141–169; Emilio Segrè, *A Mind Always in Motion: The Autobiography of Emilio Segrè* (Berkeley: University of California Press, 1993); and Spencer Weart, *Scientists in Power* (Cambridge, Mass.: Harvard University Press, 1979). Spencer Weart's *Nuclear Fear: A History of Images* (Cambridge, Mass.: Harvard University Press, 1988) deals not only with latter-day nuclear fear but also with earlier fears of worldwide destruction.

On these and other developments in military applications of science, especially nuclear science, as World War II began, see Michael Fortun and S. S. Schweber, "Scientists and the Legacy of World War II: The Case of Operations Research (OR)," *Social Studies of Science* 23 (1993): 595–642; John L. Heilbron, Robert W. Seidel, and Bruce R. Wheaton, *Lawrence and His Laboratory: Nuclear Science at Berkeley, 1931–1961* (Berkeley: Office for History of Science and Technology at University of California at Berkeley, 1981); John L. Heilbron and Robert W. Seidel, *Lawrence and His Laboratory. A History of the Lawrence Berkeley Laboratory*, vol. 1 (Berkeley: University of California Press, 1989); James G. Herschberg, *James B. Conant: Harvard to Hiroshima and the Making of the Nuclear Age* (New York: Knopf, 1993); Thomas P. Hughes, *American Genesis: A Century of Invention and Technological Enthusiasm* (New York: Penguin, 1989); Robert Jungk, *Brighter Than a Thousand Suns: A Personal History of the Atomic Scientists* (New York: Harcourt Brace Jovanovich, 1958); Carroll Pursell, "Science Agencies in World War II: The OSRD and Its Challenges," in Nathan Reingold, ed., *The Sciences in the American Context: New Perspectives* (Washington, D.C.: Smithsonian Institution Press, 1979), pp. 359–378; Richard Rhodes, *The Making of the Atomic Bomb* (New York: Simon and Schuster, 1986); and Mark Walker, *German National Socialism and the Quest for Nuclear Power, 1939–1949* (Cambridge: Cambridge University Press, 1989).

On big machines and big science, some excellent work has appeared in the last decade, including Peter Galison, *How Experiments End* (Chicago: University of Chicago Press, 1987), and Peter Galison and Bruce Hevly, eds., *Big Science: The Growth of Large-Scale Research* (Palo Alto, Calif.: Stanford University Press, 1992); and *Science after '40*, a special issue of *Osiris* (2d series, 7 [1992]) edited by Arnold Thackray. See also the issue of *Osiris* entitled *Historical Writing on American Science* (2d series, 1 [1985]), edited by Sally G. Kohlstedt and Margaret Rossiter.

I have quoted from Jules Henri Poincaré, "The Principles of Mathematical Physics," in Sopka, ed., *Physics for a New Century* (1986), pp. 281–289.

Conclusion

On science after 1940, again see the issue called *Science after '40, Osiris,* (1992) edited by Thackray. It includes James H. Capshew and Karen A. Rader, "Big Science: Price to the Present," pp. 3–25. And see Paul Forman, "Social Niche and Self-Image of the American Physicist," in Michelangelo de Maria et al., eds., *The Restructuring of Physical Sciences in Europe and the United States, 1945–1960.* (Singapore: World Scientific, 1989), pp. 96–104.

Some autobiographical books by scientists that are germane to issues discussed in the conclusion are Carl Djerassi, *The Pill, Pygmy Chimps, and Degas' Horse: The Autobiography of Carl Djerassi* (New York: Basic Books, 1992), and James D. Watson, *The Double Helix: A Personal Account of the Discovery of the Structure of DNA* (London: Weidenfeld and Nicolson, 1968). Richard Feynman's entertaining and provocative books include, *"Surely You're Joking, Mr. Feynman!": Adventures of a Curious Character* (New York: Norton, 1985) and *"What Do* ***You*** *Care What Other People Think?": Further Adventures of a Curious Character* (New York: Norton, 1988).

Karl Popper's first widely read statement on falsification was the English translation of his *The Logic of Scientific Discovery* (London: Hutchinson, 1959). For some reading on the social construction of science and the sociological analysis of claims for scientific objectivity, see Bruno Latour, *Science in Action: How to Follow Scientists and Engineers through Society* (Cambridge, Mass.: Harvard University Press, 1985); Helen Longino, *Science as Social Knowledge* (Princeton: Princeton University Press, 1990); and Andrew Pickering, ed., *Science as Practice and Culture* (Chicago: University of Chicago Press, 1990). Also, Sheila Jasanoff et al., eds., *Handbook of Science and Technology Studies* (Thousand Oaks, Calif.: Sage, 1994).

Name Index

Subject Index

The Author

Mary Jo Nye is Thomas Hart and Mary Jones Horning Professor of the Humanities and Professor of History at Oregon State University. She studied the history of science at the University of Wisconsin and taught from 1970 to 1994 at the University of Oklahoma. She is a past president of the History of Science Society and the author of numerous articles and books on the history of chemistry and physics. Her most recent book is *From Chemical Philosophy to Theoretical Chemistry: Dynamics of Matter and Dynamics of Disciplines, 1800–1950* (California, 1993).